Ethnobiology

Series editors:

Robert Voeks, Center for Remote Sensing & California State University, Fullerton, CA, USA
John Richard Stepp, Department of Anthropology, University of Florida, Gainesville, FL, USA

Ethnobiology is the study of the dynamic relationship between plants, animals, people, and the environment. Academic and applied interests include ethnobotany, ethnozoology, linguistics, paleoethnobotany, zooarchaeology, ethnoecology, and many others. The field lies at a dynamic intersection between the social and biological sciences. The major contribution from the biological sciences has come from economic botany, which has a rich historical and scientific tradition. Indeed, the objectives of the colonial enterprise were as much about the quest for "green gold"—herbal medicines, spices, novel cultivars, and others—as it was for precious metals and sources of labor. The view that ethnobiology concerns mostly the discovery of new and useful biota extended into the 20th century. The social sciences have contributed to the field in both descriptive studies but also within quantitative approaches in cognitive anthropology that have led to general principles within ethnobiological classification. Ethnobiological research in recent years has focused increasingly on problem solving and hypothesis testing by means of qualitative and especially quantitative methods. It seeks to understand how culturally relevant biotas are cognitively categorized, ranked, named, and assigned meaning. It investigates the complex strategies employed by traditional societies to manage plant and animal taxa, communities, and landscapes. It explores the degree to which local ecological knowledge promotes or undermines resource conservation, and contributes to the solution of global challenges, such as community health, nutrition, and cultural heritage. It investigates the economic value and environmental sustainability to local communities of non-timber forest products, as well as the strategies through which individual ecological knowledge and practices encourage resilience to change—modernization, climate change, and many others. Most importantly, contemporary ethnobiological research is grounded in respect for all cultures, embracing the principles of prior informed consent, benefit sharing, and general mindfulness.

More information about this series at http://www.springer.com/series/11551

John Richard Stepp

Ethnoecology and Medicinal Plants of the Highland Maya

Springer

John Richard Stepp
Department of Anthropology
University of Florida
Gainesville, FL, USA

ISSN 2365-7553 ISSN 2365-7561 (electronic)
Ethnobiology
ISBN 978-3-030-09882-7 ISBN 978-3-319-69315-6 (eBook)
https://doi.org/10.1007/978-3-319-69315-6

Softcover re-print of the Hardcover 1st edition 2018

This Springer imprint is published by the registered company Springer Nature Switzerland AG
The registered company address is: Gewerbestrasse 11, 6330 Cham, Switzerland

Preface

I first visited the Highlands of Chiapas, Mexico, over 20 years ago while traveling through Mexico and Central America. Like many before me, I was immediately drawn to the lush montane environment and its indigenous inhabitants, the Highland Maya. The Zapatista uprising had yet to take place, but the conditions that gave birth to it were on open display. The paradox of how gut-wrenching poverty could coexist with such natural beauty was not lost on me.

Several years later when the opportunity to do research in Chiapas arose, I knew that my work should be for the benefit of the Highland Maya. Around this time, the concept of biocultural diversity was gaining prominence, and it became clear that effective conservation meant not just protecting the biophysical environment but also required that language and culture be maintained as well. It was (and is) in this spirit that I undertook an investigation of how the Highland Maya interact with and know their environment and how this contributes to their health through medicinal plants.

This research has taken many twists and turns along the way, but it remains motivated by the original concerns. Despite having now worked in the area for over a decade (with extended field stays interspersed with shorter visits and check-ins with communities), it is very much a work in progress. While Chiapas is one of the most intensively explored regions in the world from both an anthropological and ethnobiological standpoint, there is still much to learn.

Change is occurring in Chiapas in ways both good and bad, but remarkably, traditional ecological knowledge persists. Five-year-old children are still likely to know dozens of plants used for food and medicines. The average adult still carries around enough plant knowledge to treat most if not all of the common ailments he/she might encounter. This is not to imply that such knowledge is static, unvarying, and never changing. Quite to the contrary, it is dynamic and variable, and this very fact is one of the reasons it persists.

The Highland Maya have endured much over the last 500 years, beginning with the European conquest and continuing to the political upheaval present in the region to this day. Looming on the horizon and making its presence increasingly known is climate change. Through it all, the language and culture has persevered while also adapting to changing circumstances. A large part of this resilience has come from the abundant local plant resources that provide most of the food and medicinal plants, both wild and cultivated. Ethnobotany (and by extension, ethnoecology) truly is the science of survival, and it is my hope that this work can play a role, however small, in conserving the rich biocultural diversity of Highland Chiapas.

Funding for this work has come from numerous sources including the National Science Foundation, Environmental Protection Agency, National Security Education Program, National Institutes of Health, National Endowment for the Humanities, The Christensen Fund, and the University of Florida Tropical Conservation and Development Program. A project of this scope involves the help and collaboration of literally thousands if one includes all of the study participants. Rather than attempt the herculean task of naming everyone and missing a few, or worse, singling out a few for praise, I offer my deepest appreciation and gratefulness to all who have been there along the way. Wokol awal!

Gainesville, FL, USA John Richard Stepp

Contents

Chapter 1
Ethnoecology and Medicinal Plants of the Highland Maya: An Introduction

Plants play a central role in human existence and medicinal plants, in particular, have allowed for the continued survival of the species. Eventually, anyone who studies human relationships to medicinal plants gets asked the question "do medicinal plants really work?" Ignoring for the time being the loaded nature of such a question, my stock response to such a question has always been "why yes they do, you're here aren't you?" This is usually followed by a quizzical look from the questioner, necessitating a more lengthy response on my part involving a broad overview of human evolution and culture, biochemical ecology, ethnobiological knowledge and practice and more. If one considers that for well over 99.9% of our species existence all we really had were medicinal plants (along with the occasional medicinal fungi, animal and even inorganic substances), they have worked remarkably well in maintaining our health and well-being.

I suppose it would be much easier to just answer yes, but its always fascinating to discover how little thought most people from western, educated, industrialized, rich and democratic (known by the acronym 'weird' cf. Henrich et al. 2010) societies give to medicinal plants-having forgotten the remedies of past generations in favor of the promise of synthetic pharmaceuticals. Plants continue to provide the base material for some of our current pharmaceuticals (Rates 2001, Stepp 2004, Newman and Cragg 2012). However, this fact has been obscured by the physical transformation of phytomedicines into pills, powders, liquids, and even suppositories. The attendant and constant advertising of new pharmaceuticals promising a cure for whatever ails you presents a narrative whereby medicines emerge not from a dynamic biophysical environment but from a sterile techno-scientific cornucopia. The explosive growth of the nutraceutical and dietary supplement industry has helped bring back some of the focus on plants but this has hardly led to a return to a nature based approach to healing in 'weird' societies.

For the non-'weird' (i.e. the rest of the world's) societies however, it is a very different story. The oft-repeated factoid (and rather egregious example of bibliographic echo) from the World Health Organization is that 80% of people in developing countries or 2/3 of the entire world's population rely on medicinal plants for

J. R. Stepp, *Ethnoecology and Medicinal Plants of the Highland Maya*, Ethnobiology, https://doi.org/10.1007/978-3-319-69315-6_1

their primary healthcare (Farnsworth et al. 1985). While this is an educated guesstimate at best, the fact remains that medicinal plants are important to a sizeable number of people on this planet.

The overwhelming importance of medicinal plants in both our past and our current world makes the understanding of human relationships to medicinal plants a crucial area of scholarship of relevance to a broad range of academic disciplines as well as policymakers and health care practitioners. I can think of no other field that has made as much of a contribution in this arena as ethnobotany, ethnobiology and ethnoecology. As noted in the Kau'i Declaration, ethnobotany is the science of survival (Prance et al. 2007) and medical ethnobotany is at the core of this equation.

The field of medical ethnobotany experienced a surge of interest starting in the 1990s (for but a few examples see Alexiades 1996; Berlin and Berlin 1996; Cox and Balick 1994; Etkin 1993; Farnsworth and Soejarto 1991; Iwu 1996; Lewis and Elvin-Lewis 1995; Martin 1995; Schultes and von Reis 1995, Moerman et al. 1999) and continued to grow into the early part of the twenty-first century with the rise of new publishing outlets such as the *Journal of Ethnobiology and Ethnomedicine*, *Ethnobiology Letters* and *Contributions in Ethnobiology* expanding the boundaries of the discipline (Pieroni et al. 2005; Nolan and Pieroni 2013; Wolverton et al. 2010; Society of Ethnobiology 2018). Meanwhile long-established journals like *Economic Botany* and *Journal of Ethnopharmacology* have continued to publish important work on medicinal plants. A more troubling trend, but indicative of growing interest in the field, is the appearance of a number of predatory journals exploiting the open access publishing system for profit with little to no attention paid to established standards of academic quality and peer-review. I will not mention them by name but sometimes their titles closely match the legitimate and established journals mentioned above.

As the field has grown, many studies have indicated that indigenous peoples have a comprehensive, empirically based knowledge of their environments (Conklin 1954; Berlin et al. 1974; Alcorn 1984; Frechione et al. 1989; Posey 1984; Moran 1982). However, there has been less research on the interaction between cultural knowledge and the use of medicinal plants in relation to the biophysical environment. One reason for the increased interest in ethnobotany is due to the exotic association of medicinal plants with primary forest. This association began to be heavily promoted in the 1990s through several popular press books and articles (e.g. Joyce 1994; Plotkin 1993).

Conservationists were quick to seize on the proposition that primary tropical rainforest contains many important medicinal plants that may hold the key to cures for diseases such as AIDS and cancer because it provides a powerful economic and medical argument for preservation of tropical rainforests. Voeks (2018) presents a compelling account of how this narrative of what he terms the "Ethnobotany of Eden" came into being. The assumption that mature forests are the most significant habitat for local peoples to obtain medicinal plants went unchallenged for many years until researchers started to explore where plants were actually being found (for a few examples, see Voeks 1996; Caniago and Siebert 1998; Stepp and Moerman 2001; Albuquerque et al. 2007). As so often happens, the reality is much more complex than what was first assumed.

This book, based on over a decade of research in Southern Mexico with the Highland Maya (specifically the Tenejapa Tzeltal Maya), explores the relationship between medicinal plants, traditional ecological knowledge and the environment. The biodiversity of the region remains among the highest in the world, comprising more than 9000 plant species (Breedlove 1981, 1986; Toledo 1988; Ramamoorthy et al. 1993). Over 1600 plants are employed for medicinal uses. Knowledge for approximately 600 species is widespread, while there is a large consensus on around a couple hundred or so plant species (Stepp 2017; Berlin and Berlin 1996). Medicinal plants play an overwhelmingly primary role in the daily health care of the Highland Maya.

This book is an ethnoecological study of medicinal plants of the Highland Maya in Chiapas and it has three main objectives: (1) identifying which medicinal plants are actually used; (2) determining the role of environmental variation on use and selection of medicinal plants; and (3) identifying which habitats are preferred for medicinal plant procurement. This study is based primarily on ethnographic research involving extended fieldwork over the last two decades and direct interaction and observation. Multiple methods, both quantitative and qualitative, were utilized. The research design is systematic as opposed to interpretive and for the most part involves an explanatory approach. "Explanatory approaches generally involve testing elements of theory that may already have been proposed in the literature or that have been informed by exploratory research. Research designs in this mode are determined *a priori* and their primary purpose is to eliminate threats to validity, where validity is concerned with whether things are what they appear to be or are the best approximation to the truth. In this enterprise, explanation can involve a general search for causality or prediction" (Johnson 1998:139). The specific methodologies and research techniques are more fully addressed in subsequent chapters.

Both the general ethnobotany (Berlin et al. 1974) and the medical ethnobotany (Berlin and Berlin 1996) are well understood. This means that the Highland Maya are one of the few indigenous populations in the world with which a study such as this one could be undertaken without first investing significant amounts of time in obtaining baseline ethnobotanical data. Within the last few decades there has been increased environmental degradation and population growth in Chiapas, Mexico. However, the biodiversity of the state of Chiapas remains among the highest in the world. The Highland Maya medical system is heavily reliant on local medicinal plant species and is integrally interwoven with the biophysical environment. By building on this previous research it becomes possible to relate this botanical knowledge (ethnobotany) to ecological knowledge (ethnoecology). Ethnoecology is an interdisciplinary field that explores indigenous cognitive perceptions of divisions in the biological world and how human behavior relates to these perceptions (Frechione et al. 1989). As such, this work has two interrelated domains of investigation: cognitive and behavioral. This approach allows for a fuller understanding of human-environmental relations by contextualizing human behavior within local perceptions of the environments.

The affordance concept laid out by Gibson (1979) in particular, provides a useful framework to look at healing within an environmental context. "The affordances of the environment are what it offers the animal, what it provides or furnishes, either for good or ill…it implies the complementarity of the animal and the environment (Gibson 1979:127)." The utility of the concept lies in its emphasis on the holistic aspects of human-environment relations and the complementary nature of this relation, in this case between medicinal plants and humans. Inherent in the definition are the values and meanings attached to objects in nature. Additionally it implies not just the perceptual features of objects in the environment but biological features (in the case of medicinal plants, pharmacological and toxic aspects) and cultural features.

1.1 Generalized Versus Specialized Knowledge

A major premise of this work is that ethnobotanical knowledge among the Highland Maya is common, mundane, widespread and generalized. As I once said elsewhere, "asking a Highland Maya what medicinal plants they use is no different from asking them what they had for breakfast." (Stepp 2002:791). While this approach has been validated through numerous studies of other cultures (e.g. Conklin 1954; Alcorn 1984) and especially in Chiapas (Berlin et al. 1974; Berlin and Berlin 1996) there has been resistance to such an approach by ethnobotanists that tend to work exclusively with specialized healers (cf. the discussion between Berlin, Balick, Cox, Elvin-Lewis, Lozoya, Martin, Iwu and Posey in Berlin and Berlin 1994:260–61) and this divide continues to this very day. Many medical ethnobotanical studies continue to rely exclusively on interviews with specialized healers. Stereotypically, this approach is much more common with ethnobotanical studies done by researchers with a background in botany while those with anthropological backgrounds more often investigate generalized knowledge. The reasons for this dichotomy have much to do with the development of the field of ethnobotany within two larger disciplines that often did not communicate across disciplinary boundaries. But apart from historical aspects of ethnobotany there are other possible reasons why the ethnobotany of specialized healers has remained a popular approach. In western culture medical knowledge has become increasingly specialized and even sacred (Laurel 2018; Katz 1999). Despite efforts at objectivity, cultural patterning influences research design. Given the lack of self-treatment in western industrialized societies and the overwhelming importance assigned to specialized practitioners of medicine, it is perhaps not surprising that so much research has focused on specialized healing in other cultures.

1.2 Why Study the Ethnoecology of Medicinal Plants?

In a general sense, ethnoecological studies can contribute to a bridging of adaptational (materialist) and ideational theoretical approaches in anthropology (cf. Alves and Albuquerque 2016; Hunn 1989) by looking at both knowledge and behavior. This book adds to this growing body of literature and addresses a significant gap by looking at traditional knowledge and use of medicinal plants within an ethnoecological framework. Despite decades of research on medical ethnobotany, the systematic study of *actual* medicinal plant use over time remains underexplored. The study of medicinal plants as a domain in ethnoecology can provide a greater understanding of plant-human interactions and human-biophysical environments.

There are significant applied aspects in this research. Medicinal plant conservation is a relatively recent endeavor that has gained prominence only in the last 20 years or so (Sheldon et al. 1997). Today a third of all plant species are at risk of extinction (Pimm and Joppa 2015). This does not include the number of plants in danger of local extirpation, a problem that has widespread implications for local peoples reliant on traditional plant medicine for their primary health care. This book provides data that could be used to suggest improved management of the resource in the context of increasing human population pressures in the Highlands (INEGI 2011). It also has the potential to make recommendations to governmental and nongovernmental organizations working on sustainable resource initiatives in Mexico and beyond. On a basic inventory level, besides having importance for other ethnobotanical research within Chiapas, the research potentially relates to ethnobotanical research throughout the neotropics. Many of the plants utilized by the Highland Maya are also used by other sociolinguistic groups in Mexico, Central America and beyond (cf. Frei et al. 2000; Balick and Arvigo 2015). Because of the overwhelming significance of plants in healing practices for the vast majority of the world's population, it is crucial to understand the cultural relationships underlying these resources. This can help to ensure the continued survival of traditional peoples and their healing practices.

Understanding the role of disturbed areas and medicinal plants can contribute to theoretical developments regarding human modification of environments, a matter becoming more pressing by the day as the world realizes that we have entered the anthropocene (Steffen et al. 2007). We live in a human modified world: a human ecosystem (Abel and Stepp 2003) and effective conservation requires being to attuned to this reality and incorporating traditional ecological knowledge and human cultural diversity into the framework (Gavin et al. 2015).

1.3 Prior Informed Consent Procedures

The process of prior informed consent (PIC) is crucial in all research involving human participants. It is also a process fraught with difficulty for a variety of reasons. Chief among them are defining what it means to give informed consent and determining at which levels it is necessary to obtain consent and from whom. Standards and definitions of PIC vary widely and are often unsatisfactory within the context of anthropological research. The International Society of Ethnobiology has explored this issue extensively and has a strong (and evolving) code of ethics <http://www.ethnobiology.net/what-we-do/core-programs/ise-ethics-program/code-of-ethics/>. This research follows the guidelines provided therein. Ultimately, the responsibility lies with the researcher to ensure that the well being of the people participating in the study takes precedence over any other concerns.

I outline below the ongoing steps that I have taken in this research to obtain PIC. These steps not only conform to international expectations but, most importantly, to community and regional expectations. The first step in the process is to obtain written permission from the municipal authorities to conduct research in the municipality. A presentation of the research to be carried out is made to the municipal president. The president then provides a letter of agreement which grants permission to conduct research in the municipality at large. In actuality, this letter only gives one permission to *ask permission* because it is at the community level where real consent is obtained. Individual communities have a great deal of autonomy in controlling who conducts research within their boundaries and are in no way bound to permit research simply because an investigator has permission at the municipality level.

At the community level, the permission process varies slightly from community to community. After identifying potential communities where I want to work, I visit the community. This is usually done in the late afternoon when we were most likely to find community authorities at home, having often just returned from working in their milpas or coffee groves. I would ask around and find a *comité* (literally "committee" but in the Spanish spoken by the Highland Maya in Chiapas it also refers to the head of the committee; oftentimes the various committees would consist of but a single individual). *Comités* in Highland Maya communities serve a variety of political and cultural functions. There are committees devoted to education, health, municipal relations, agriculture, cargoes, etc. Committees can be formed *ad hoc* depending on the needs of the community.

After identifying a committee member, I describe the project in some detail and my collaborators translate into Tzeltal Maya. I then ask when we could meet with the entire assembly of *comités* to further explain the project. Sometimes the *comité* would tell us to come back (a matter of asking permission to ask permission) or we would be given a date and time where we could make our case before the *comités*.

At this next meeting, I fully describe the project (in both Spanish and Tzeltal): explaining the goals, how the knowledge obtained would be utilized, how the participants would be compensated and that the community could at any time terminate our presence. This process of explanation often goes on for more than an hour as the *comites* are keen to ask questions in order to understand the project fully. Because so much anthropological fieldwork has been conducted in the region, people are generally knowledgeable about the nature of anthropological fieldwork and dissertation research.

At this point, if approval is obtained I would then be invited to present the project once again at a full assembly of the community. These *asembleas generales*, despite the somewhat marginalized position afforded to women at them, are highly democratic and consensus oriented. While the process seems somewhat straightforward, if not a little laborious, there are sometimes additional steps needed before the full community was presented with the project. It also important to note that all plant species described herein have been previously noted for their medicinal uses and are thus considered prior art of the Highland Maya, not subject to patenting and form part of a medicinal plant commons.

1.4 Organization of the Book

The book is organized as follows:

Chapter 1 (this chapter) provides an introduction and overview of the research.

Chapter 2: A multiple environments framework is used to explore the biophysical and sociocultural environments of the Highland Maya. Biocultural diversity in Chiapas with a focus on the Highlands is described along with ecological, economic, demographic geographic, geological, historical and cultural aspects. A detailed overview of previous ethnobiological and ethnographic work in the region is presented. Research communities in the municipality of Tenejapa are explored in detail.

Chapter 3: Illness in Highland Chiapas is far too common due to a number of historical and environmental factors. Gastrointestinal and respiratory diseases account for 87% of illnesses. Illness increases significantly at the onset of the summer rainy season. The vast majority of illnesses are naturalistic in etiology, empirically diagnosed and self-treated. Visits to specialized healers are not very common and are usually done specifically to treat illnesses with personalistic etiologies. Through use recall interviews, actual plant use was determined. Management, selection, time allocation and ecological data were also collected. Most medicinal plants are collected within a short distance of the household. The Tzeltal Maya ethnoecological classification system is presented and used to examine medicinal plant habitats. The vast majority of medicinal plants are procured in human modified habitats and not forested areas. Weeds are disproportionately represented in the medicinal flora. A comparison is made between where plants are actually collected and where plants are known to grow.

Chapter 4: A botanical ethnography is presented for the most important medicinal plants used by the Highland Maya in the municipality of Tenejapa. Importance is determined based on use reports over time and ethnographic interviews. Despite knowing approximately 1500 medicinal plants, five plant species account for around 25% of all uses. Linguistic, economic and ecological aspects of medical ethnobotany are described. In addition, a reported ethnoecological distribution of these medicinal plants based on degree of human modification of the landscape and gross climatic zones is presented. Plants are listed in order of their importance. For each plant the entry is divided into two sections: Ecology, Botany, Folk Classification and Nomenclature; and Medical Ethnobotany and Ethnopharmacology.

Chapter 5: Economic arguments have been made to justify conservation of tropical mature forests. These forests are assumed to be rich in potential pharmaceuticals and medicinal plants. Conservationists and ethnobotanists have made this point in a number of fora. A related assumption is that mature tropical forests are a significant habitat for indigenous peoples to find medicinal plants. Systematic studies of *actual* plant use and procurement have been rare in ethnobotany. However, there are several studies that suggest mature tropical forest is not significant habitat for indigenous peoples to obtain medicinal plants. In Highland Chiapas, we found no use of mature forest as a source of medicinal plants. Young secondary forest is rarely utilized to obtain medicinal plants. The same pattern is observed in communities that are adjacent to mature forests and could have easy access to them.

Chapter 6: A theoretical basis is provided for the main findings, drawing from human ecology and biological ecology. Highland Maya overwhelmingly favor human modified environments for medicinal plants. Explanatory factors related to human ecology are discussed. Medicinal plants need to be abundant and accessible and plants that are close by are preferred. A medicinal flora also needs to be efficacious and biologically active. Ecological and biochemical theories and findings in relation to distribution and concentration of plant secondary compounds explain why plants from disturbed areas would tend to be more bioactive.

Chapter 7: The implications of this work for conservation, health, and the environment is discussed. The potential impact of climate change on medicinal plant ethnoecology in Highland Chiapas is examined. Climate change is already occurring in Highland Chiapas. The most utilized medicinal plant species will likely experience little impact from climate change due to their widespread availability and abundance. However, more obscure medicinal plant species are likely to be impacted. In addition to affecting medicinal plant yield, climate change can have an impact on medicinal plant quality. Future directions in medical ethnobotanical research in the region are explored.

References

Abel T, Stepp JR (2003) A new ecosystems ecology for anthropology. Conserv Ecol 7(3):11

Albuquerque UP, de Medeiros PM, de Almeida AL, Monteiro JM, Neto EM, de Melo JG, dos Santos JP (2007) Medicinal plants of the Caatinga (semi-arid) vegetation of NE Brazil: a quantitative approach. J Ethnopharmacol 114(3):325–354

Alcorn JB (1984) Huastec Maya ethnobotany. University of Texas Press, Austin

Alexiades M (ed) (1996) Selected guidelines for ethnobotanical research: a field manual. NYBG Press, New York

Alves AGC, Albuquerque UP (2016) Ethnobiology or ethnoecology? In: Albuquerque U, Nóbrega Alves R (eds) Introduction to ethnobiology. Springer, New York

Balick MJ, Arvigo R (2015) Messages from the gods: a guide to the useful plants of Belize. Oxford University Press, New York

Berlin B, Berlin EA (1994) Anthropological issues in medical ethnobotany. In: Chadwick DJ, Marsh J (eds) Ethnobotany and the search for new drugs. Wiley, Chichester

Berlin EA, Berlin B (1996) Medical ethnobiology of the Highland Maya of Chiapas, Mexico: the gastrointestinal diseases. Princeton University Press, Princeton

Berlin B, Breedlove DE, Raven P (1974) Principles of Tzeltal plant classification. An introduction to the botanical ethnography of a Mayan-speaking people of the highlands of Chiapas. Academic, New York

Breedlove DE (1981) Introduction to the flora of Chiapas. In: Breedlove DE (ed) Flora of Chiapas, part 1. California Academy of Sciences, San Francisco

Breedlove DE (1986) Listados florísticos de México. IV. Flora de Chiapas. Instituto de Biología, UNAM, México Federal

Caniago I, Siebert SF (1998) Medicinal plant ecology, knowledge and conservation in Kalimantan, Indonesia. Econ Bot 52(1):229–250

Conklin HC (1954) An ethnoecological approach to shifting agriculture. Trans N Y Acad Sci Series II 17:133–142

Cox PA, Balick MJ (1994) The ethnobotanical approach to drug discovery. Sci Am 270:82–87

Etkin N (1993) Anthropological methods in ethnopharmacology. J Ethnopharmacol 38:93–104

Farnsworth NR, Soejarto DD (1991) Global importance of medicinal plants. In: Akerele O, Heywood V, Synge H (eds) The conservation of medicinal plants. Cambridge University Press, Cambridge

Farnsworth N, Akerele O, Bingel AS (1985) Medicinal plants in therapy. Bull World Health Organ 63:965–981

Frechione JD, Posey DA, Da Silva LF (1989) The perception of ecological zones and natural resources in the Brazilian Amazon: an ethnoecology of Lake Coari. In: Advances in economic botany, vol 7. New York Botanical Garden, New York, pp 260–282

Frei B, Sticher O, Heinrich M (2000) Zapotec and Mixe use of tropical habitats for securing medicinal plants in Mexico. Econ Bot 54:73–81

Gavin MC, McCarter J, Mead A, Berkes F, Stepp JR, Peterson D, Tang R (2015) Defining biocultural approaches to conservation. Trends Ecol Evol 30(3):140–145

Gibson J (1979) The ecological approach to visual perception. Lawrence Erlbaum, New York

Henrich J, Heine SJ, Norenzayan A (2010) Beyond WEIRD: towards a broad-based behavioral science. Behav Brain Sci 33(2–3):111–135

Hunn E (1989) Ethnoecology: the relevance of cognitive anthropology for human ecology. In: Morris WFB (ed) The relevance of culture. Bergin and Garvey Publishers, New York

INEGI (2011) Mexico population and housing census 2010. INEGI, Aguascalientes

Iwu M (1996) Implementing the biodiversity treaty: how to make international cooperative agreements work. Trends Biotechnol 14:78–83

Johnson J (1998) Research design and research strategies. In: Bernard HR (ed) The handbook of methods in cultural anthropology. Altamira Press, Walnut Creek

Joyce C (1994) Earthly goods: medicine hunting in the rainforest. Little Brown, New York

Katz P (1999) The scalpel's edge: the culture of surgeons. Allyn and Bacon, Boston

Laurel SJD (2018) Medicine as a sacred vocation. Proc (Baylor Univ Med Cent) 31(1):126–131

Lewis WH, Elvin-Lewis M (1995) Medicinal plants as sources of new therapeutics. Ann Mo Bot Gard 82:16–24

Martin G (1995) Ethnobotany: a methods manual. Chapman Hall, New York

Moerman DE, Pemberton RW, Kiefer D, Berlin B (1999) A comparative analysis of five medicinal floras. J Ethnobiol 19(1):49–67

Moran E (1982) Developing the Amazon. Indiana University Press, Bloomington

Newman DJ, Cragg GM (2012) Natural products as sources of new drugs over the 30 years from 1981 to 2010. J Nat Prod 75(3):311–335

Nolan J, Pieroni A (2013) Recollections, reflections, and revelations: ethnobiologists and their "first time" in the field. J Ethnobiol Ethnomed 9:12

Pieroni A, Price L, Vandebroek I (2005) Welcome to journal of ethnobiology and ethnomedicine. J Ethnobiol Ethnomed 1:1

Pimm SL, Joppa LN (2015) How many plant species are there, where are they, and at what rate are they going extinct? Ann Mo Bot Gard 100(3):170–176
Plotkin MJ (1993) Tales of a Shaman's apprentice. Viking Press, New York
Posey DA (1984) A preliminary report on diversified management of tropical forest by the Kayapó Indians of the Brazilian Amazon. In: Prance GT, Kallunki JA (eds) Ethnobotany in the neotropics. New York Botanical Garden, New York
Prance G et al (2007) Ethnobotany, the science of survival: a declaration from Kaua'i. Econ Bot 61(1):1–2
Ramamoorthy TP, Bye R, Lot A, Fa J (1993) Biological diversity of Mexico. Oxford University Press, Oxford
Rates SMK (2001) Plants as sources of drugs. Toxicon 39:603–613
Schultes RE, von Reis S (1995) Ethnobotany: evolution of a discipline. Timber Press, Portland
Sheldon JW, Balick MJ, Laird SA (1997) Medicinal plants: can utilization and conservation coexist. Advances in economic botany, vol 12. New York Botanical Garden, Bronx
Society of Ethnobiology (2018) https://ethnobiology.org/publications/contributions. Accessed 14 Jan 2018
Steffen W, Crutzen PJ, McNeill JR (2007) The anthropocene: are humans now overwhelming the great forces of nature. AMBIO J Hum Environ 36(8):614–621
Stepp JR (2002) On Maya medicine and the biomedical gaze. Curr Anthropol 43(5):791–792
Stepp JR (2004) The role of weeds as sources of pharmaceuticals. J Ethnopharmacol 92:163–166
Stepp JR (2017) Transmission, sharing and variation of medicinal plant knowledge and implications for health. In: Olson L, Stepp JR (eds) Plants & health: new perspectives on the human relationships to medicinal plants. Springer, New York, pp 163–170
Stepp JR, Moerman DE (2001) The importance of weeds in ethnopharmacology. J Ethnopharmacol 75:19–23
Toledo V (1988) La Diversidad Biológica de México. Ciencia y Desarrollo 181:17–30
Voeks R (1996) Tropical forest healers and habitat preference. Econ Bot 50:381–400
Voeks R (2018) The ethnobotany of Eden: rethinking the jungle medicine narrative. University of Chicago Press, Chicago
Wolverton S, Fowler C, Cozzo D (2010) Introducing ethnobiology letters. Ethnobiol Lett 1:1–2

Chapter 2
Environmental Context of the Highland Maya

The term 'environment' is sometimes used in an imprecise and reified manner in ethnoecology and ethnobiology. "The environment" is taken to mean only the biophysical environment; the earth system; or in a loose colloquial sense, 'nature.' Using the term in this manner obscures the relationship between environment and agent and how this interaction forms a system or ecosystem. In the interest then, of a more precise usage of the term, I present the following distinction. Using a multiple environments approach (Stepp 1999; Kuchka 2001), 'environment' is taken to mean, at a minimum, the biological, physical, social, and cultural environments. When a more precise usage is intended one can then refer to the specific environmental domain. One way to depict the taxonomy of these multiple environments is with a model of concentric circles arranged in an evolutionary sequence as in Fig. 2.1. In this model, a physical environment exists prior to biological forms, some of which develop sociality and culture. Also implicit in this arrangement is the concept of a system being composed of both an input environment and a output environment and its associated network relationships (Schizas and Stamou 2017; Patten 1978). The lines represent matter, energy and information flows. Ecosystems form through the dynamic coupling of one agent's output environment with another agent's input environment (cf. Patten 1978; Odum 1993).

An ecological perspective then is one that takes these environments into account and looks at the interactions between them.

2.1 A Note About Sources of Data and Their Accuracy

Much of the data presented in this chapter concerning demographic, economic and physiographic conditions comes from various reports provided by the Mexican federal government through the Instituto Nacional de Estadística, Geografía e Informática (INEGI). INEGI is in charge of collecting census information for the entire country along with production of maps with various themes, INEGI has become increasingly

J. R. Stepp, *Ethnoecology and Medicinal Plants of the Highland Maya*,
Ethnobiology, https://doi.org/10.1007/978-3-319-69315-6_2

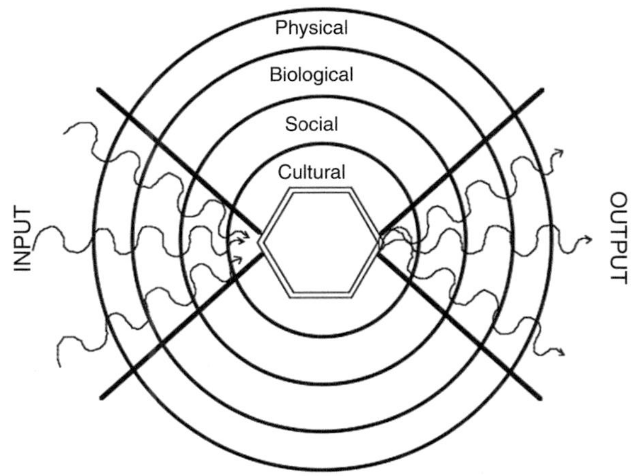

Fig. 2.1 A hierarchy of multiple environments (after Stepp 1999; Kuchka 2001)

more sophisticated in recent years and they now feature much of their data available directly from their website (http://www.inegi.gob.mx). INEGI also has stores in most of the capitals of Mexican states and in the Mexico City airport where one can purchase a wide range of topographical maps, satellite images and books.

The services offered by INEGI provide invaluable data for researchers. However, researchers sometimes question the validity of some of the data that INEGI provides (e.g. Berlin and Berlin 1996). At times, the level of detail that they provide for highly rural municipalities such as the ones in the Highlands of Chiapas is astounding. It seems almost impossible to collect some of this data without a legion of researchers tromping across the Highlands. During my years of fieldwork I have encountered numerous visitors to rural communities such as social service workers, other anthropologists, military personnel, peace activists, tourists, Coca-Cola deliverymen and even a few Jehovah's Witness missionaries. But I have never run into anyone from INEGI in Highland Maya communities. Because INEGI researchers are rarely, if ever seen, in the Highlands, this had led to some questions about their methodology and validity for some of their reports.

I did not set out to verify the INEGI data so I can only provide a limited report of my own experience with INEGI data. For example, I use INEGI topographical maps extensively in my research. I found that with regards to physical features and accurate reporting of altitude and contour lines, the maps are highly accurate. Unfortunately this does not extend to accuracy of place names and locations of communities. For example, the topographical map that contains most of the municipality of Tenejapa in the Highlands of Chiapas (the sheet is named Oxchuc,E152D52) places the community of Chixtontic over 1.5 km away from where it actually is. The location of the *cabecera* (or municipal center) of Tenejapa is listed twice: one is accurate, the other is across the highway from the *cabecera* for the municipality of Oxchuc.

Many of the areas that are noted as forested on the map are not. Two of the largest tracts of primary forest near the communities of Matzab and Achlum in Tenejapa are not noted as being forested land at all. While the major roads are accurately depicted on the map, many of the smaller ones are either not included or the course they follow deviates substantially from the topographical map. It appears that some of the lines that are marked as roads are, in fact, well used walking trails in the mountains. I have found the topographical features and locations of geological and hydrological features to be accurate when checked against a handheld GPS unit.

With the above in mind, the reader is cautioned about overly relying on data reported by INEGI absent any corroborating evidence or ground truthing.

2.2 Biocultural Diversity in the State of Chiapas

Chiapas, with an area of 73,887km^2, lies at the southernmost extent of Mexico and borders Guatemala on its eastern/southeastern side (Figs. 2.2 and 2.3). Chiapas is notable for its high biocultural diversity in a country that itself ranks very high in biocultural diversity by several measures. With regards to biological diversity, Mexico is considered to be one of the six megadiversity countries of the world. It ranks third, following Brazil and Colombia (Mittermeier 1988). Within the North American/Central American bioregion, Mexico ranks first in species richness and sixth worldwide for species endemism (Sisk et al. 1994). More than 17% of Mexico's 2500 genera and 52% of its 22,000 vascular species are endemic to the country (Rzedowski 1993, 1991). At present, there are at least 58 main groups of indigenous peoples speaking 282 languages and dialects in Mexico (Simons and Fenning 2018). It is ranked ninth in the world for biocultural diversity (Ryan 1993). Mexico is also considered one of the three primary centers of agricultural origin (Harlan 1975). Several of the world's most important agricultural crops, including maize, cacao, tomato, several bean species and likely capsicum peppers were domesticated there (Yamakake et al. 2009; Kraft et al. 2014).

Fig. 2.2 Political Map of Mexico

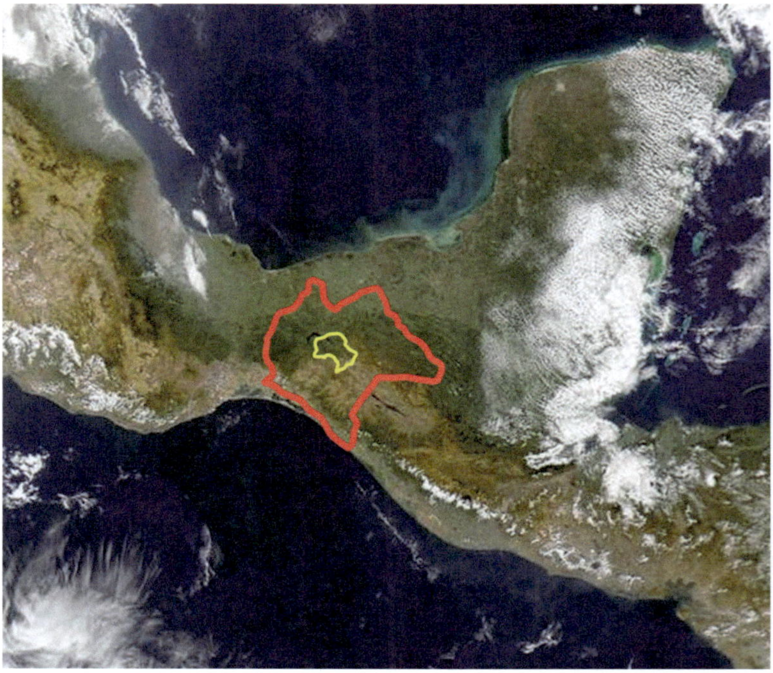

Fig. 2.3 OrbView-2 satellite image of Southern Mexico with the borders of Chiapas in red and the highland region in yellow

Within this context is the state of Chiapas as one of the two most biodiverse regions of the country along with the neighboring state of Oaxaca (Toledo 1988, 1994). In the Barthlott et al. classification of the world's diversity and endemism of vascular plants, Chiapas is given a ranking of 9 (10 possible) with more than 5000 species/10,000 km^2 (Barthlott et al. 1996). Chiapas has more than 9000 species of vascular plants and more than 1150 species of vertebrates (Breedlove 1981, 1986; Toledo 1988; Ramamoorthy et al. 1993). There are four major Maya languages spoken in Chiapas (Tzeltal, Tzotzil, Ch'ol, Tojolobal). Other Mayan languages include Lacandon, Kanjobal, and Mam. There is also a sizable population of Zoque speakers. In total, there are at least 54 indigenous languages spoken in Chiapas, although of these only 11 have more than 100 speakers (INEGI 2011). Attesting to the high cultural diversity in Chiapas is the fact that half of the ethnographic literature on indigenous people in Mexico written between 1965 and 1995 deals with indigenous people in Chiapas (Köhler 2000). The two largest sociolinguistic groups are the Tzeltal and Tzotzil Maya, living in 14 municipalities in the Highlands (Fig. 2.4) although outmigration beginning in the late 1960s has led to substantial populations moving to lowland regions of Chiapas and to other areas throughout Mexico. The Tzotzil municipality of Chamula in particular has experienced massive migration and I have encountered Chamulans in the city of Mérida and as far north as Chihuahua. As of the most recent comprehensive Mexican Census in 2010, the population of Tzeltal speakers was at least 445,856 and the Tzotzil population was at

Fig. 2.4 Map of the highland region of Chiapas, Mexico, Tzeltal communities are to the east of the dark line and Tzotzil communities lie to the west

least 404,704 (INEGI 2011). Tzeltal and Tzotzil are 2 of the 31 extant Maya languages spoken in Mexico and Central America and are part of the Greater Tzeltalan grouping that include Chol, Chontal, Chortí and Choltí (Campbell and Kaufman 1985; Law 2013).

Chiapas began to experience rapid population growth in the early 1970s. For example, the state's population almost tripled from 1.5 million in 1970 to 3.9 million in 1990. The current population is 5.4 million (INEGI 2017).

2.3 Highlands and the Municipality of Tenejapa

The specific area in the Highlands where this research took place was in the Tzeltal Maya municipality of Tenejapa. This municipality, with an area of 99.4 km^2, lies approximately 27 km to the north east of the colonial and now tourist town of San Cristóbal de las Casas along an all-weather road. Tenejapa has been the site of a great deal of anthropological fieldwork over the last 40 years. There have been numerous research articles, book chapters, monographs and dissertations as a result. To varying degree, this body of literature provides a comprehensive and accurate portrait of Tenejapa, at least up to the date in which it took place. The first published anthropological research in Tenejapa was in 1901 by Frederic Starr who made a brief visit to the municipality and collected biometric data from the population (Starr 1902). Since the late 1950s there has been a steady stream of anthropologists doing research in Tenejapa. The surge of interest in Tenejapa and other highland Maya municipalities began with the Man and Nature program under the direction of Norman McQuown at the University of Chicago. This was followed a few years later by the Harvard Chiapas Project under the direction of Evon Vogt which focused mostly on Tzotzil municipalities. McQuown wanted to place both a linguist and ethnographer in the Highland municipalities. In 1961, he recruited Brent Berlin and Andrés Medina to conduct research in Tenejapa (Berlin 1962; Medina 1991).

Researchers returned many times in the ensuing years for fieldwork. For example, Stross (1970) on childhood acquisition of botanical knowledge; Branstetter (1974) on the social significance of clothing; Hunn (1977) on Tzeltal folk zoology; Brown (1979) on politeness and gender roles; Haehl (1980) on Tzeltal kinship analysis; and Rostas (1986) on change, identity and religious practices, Maffi (1994) on ethnosymptomatology, Castille (1996) on mental illness, Luber (2002) on an ethnomedical condition known as *chalam ts'ots*, Casagrande (2002) on cognitive prototypes and medicinal plants, Lampman (2004) on ethnomycology, Adams (2004) on ethnophysiology, Brown (2007) on sociolinguistics, Doane (2011) on coffee production, Page-Pliego (2013) on diabetes and nutrition and D'Alesandro and Cabañas (2017) on nature cosmology to name just a few! The fact is that Tenejapa and Highland Chiapas are one of the most intensively studied regions in the world from a social science perspective and I encourage readers fascinated by this region to consult these sources to learn more. Rather than attempt to duplicate the work of others then, this chapter provides a description of characteristics that are particular relevant to my research and may not have been mentioned by other scholars. Moreover, this chapter does not seek to place Tenejapa within that fictive time frame known as the "ethnographic present" or the "ethnobotanical present" (Stepp 2016) but rather portrays Tenejapa of recent years with the understanding that it continues to evolve and change. Use of the present tense in this discussion then is a matter of convenience and not an attempt at what has been termed 'discursive distancing' (Fabian 1983).

2.3.1 Population and Sociopolitical Organization

The population of Tenejapa grew dramatically over the last 60 years (Fig. 2.5) and numbered 40,268 as of the most recent decadal Mexican Census (INEGI 2011). However, the rate of increase in recent years has subsided dramatically compared to the period between 1970 and 2000. The municipality is almost entirely Tzeltal Maya with only a few hundred non-indigenous people, who live mostly in the Tenejapa town center, *lum* (literally 'land') in Tzeltal or *cabecera* (literally 'head') in Spanish. Among the population aged 5 years and older, 55% of males speak Tzeltal and Spanish, and 26% of females speak Tzeltal and Spanish. Overall, 58% of Tenejapans are monolingual Tzeltal speakers. There are 5643 households in Tenejapa with an average of 5.89 people per household (INEGI 2011).

The economic and political heart of Tenejapa lies in the Tenejapa town center, a remnant of the 'vacant town' described by Tax (1937). In this model, settlement is

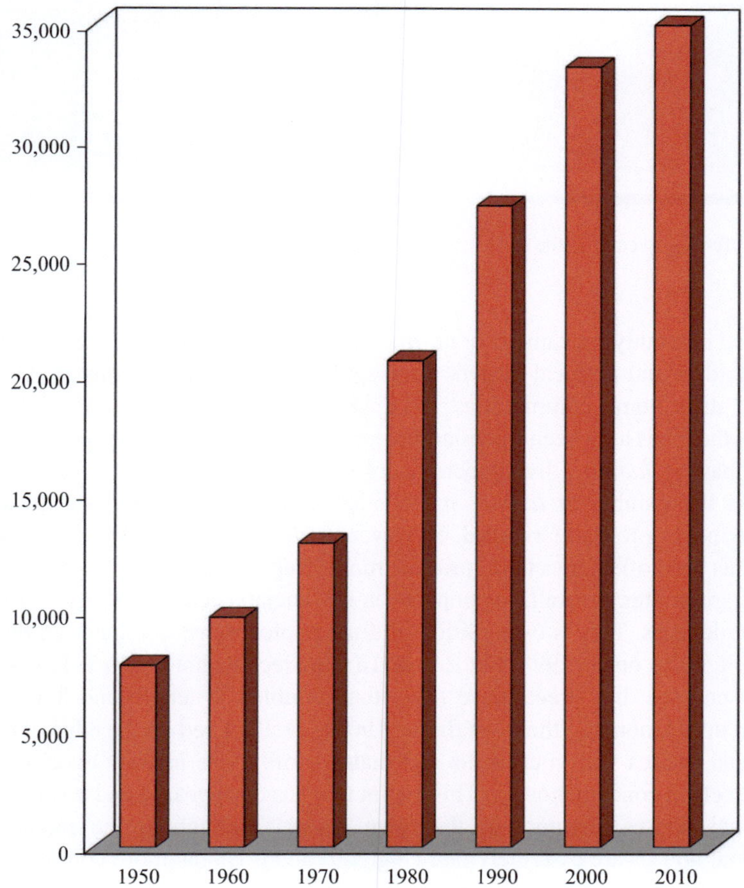

Fig. 2.5 Population of Tenejapa, Chiapas 1950–2010 (INEGI 2011)

Fig. 2.6 Tenejapa center in the early 1960s

dispersed into outlying hamlets or *parajes* and the center is occupied temporarily by cargo holders and political leaders who regularly rotate out of living in the center back to their home communities. Tenejapa never quite fit this model however because there has long been a substantial presence of non-indigenous people (known in Chiapas as *ladinos*) living permanently in Tenejapa center. During the last 30 years the number of *ladinos* in Tenejapa center has declined due to a trend towards moving to San Cristóbal. However the loss of *ladinos* living in Tenejapa center has been offset by a trend towards more Tzeltal choosing to live permanently in Tenejapa center. Overall, the population of Tenejapa center has increased in the last few decades. This is exemplified in three photographs: one taken in the early 1960s (Fig. 2.6), one in 1998 (Fig. 2.7) and a more recent photo from 2012 (Fig. 2.8). As you can see, the streets have more than doubled in length and houses have cropped up surrounding the town. In the 1960s, Tzeltal lived on the edge of town in traditional houses with mud walls and thatch roofs while *ladinos* lived along the main streets in concrete houses. This distinction has disappeared and no one lives in traditional houses anymore in the town center. Throughout the municipality, traditional houses are extremely rare (Fig. 2.9) and it is now more common to find a house with wood plank walls and tin roofs (Fig. 2.10).

Fig. 2.7 Tenejapa center in 1998

Fig. 2.8 Tenejapa center in 2012

Outside of the town center, people continue to live in dispersed settlements organized around small communities (*parajes*). These *parajes* range in populations from 50 to upwards of 2000 and usually contain some sort of civic complex including a school, administrative buildings and a basketball court. The number of *parajes* has grown from 21 in the late 1960s (Berlin et al. 1974) to 42 in the 1990 (INEGI 1993)

Fig. 2.9 Traditional style of housing in Tenejapa, community of Winik Ton 1986

Fig. 2.10 Contemporary style of housing in Tenejapa

and is now 66 (INEGI 2015). The aftermath of the Zapatista uprising in 1994 exacerbated this fissioning process due to conflict between supporters of the *organizacíon* (as the EZLN is called by Tenejapans) and those Tzeltal opposed to the EZLN. The exact processes by which *parajes* are formed and boundaries are delineated is unclear and ethnographic research is sorely needed.

Fig. 2.11 Memorial along main road in Chenalhó for Tzotzil Maya slain during the Acteal massacre

Like many Highland municipalities, political tensions run high among the population and there is frequent conflict. It is not unheard of for the state police to occupy Tenejapa center and stay for a few weeks until tensions subside. There is general mistrust in the office of the municipal presidency and it is common to find a president being removed from office well before his elected term expires. Although the PRI party has traditionally dominated politics in Tenejapa (and the rest of the country) their hold has slipped somewhat in recent years. It is common to see signs and slogans for opposition parties in Tenejapa center. Despite conflict and political violence there have not been paramilitary groups with widespread operations in Tenejapa although their presence is well documented in neighboring municipalities. The internationally known Acteal massacre on December 22, 1997 where 45 Tzotzil Maya were killed by *Paz y Justicia* paramilitaries took place only a few kilometers from the community of Majosik' in Northern Tenejapa (Fig. 2.11). In 2011, four familes of Zapatista sympathizers were attacked by PRI supporters in the community of Banavil, Tenejapa and were either killed or displaced from their homes. Tenejapa does not appear to have communities involved in autonomy move-

Fig. 2.12 Sign for autonomous community of Polho

ments, unlike other neighboring municipalities such as Chenalho (Fig. 2.12). The underlying causes for political conflict in the Highlands are unequal distribution of resources and a shortage of arable land. Despite a massive influx of development aid that has flooded the Highlands over he last 20 years, it is unlikely that these conditions will improve in the near future and conflict will likely continue.

2.3.2 *Economy*

Despite widespread changes that have occurred in Tenejapa, the vast majority of Tenejapans are still engaged in the same economic activity that they have been doing for millennia: subsistence farming. The major crop is maize, supplemented by several types of beans and squashes and a variety of other crops depending on the altitude. At lower elevations below 1800 m, coffee is common. There is often conflict with the coffee growers and the *coyotes*, or middlemen, that take the coffee from the highlands to regional processing centers. Coffee prices in Tenejapa have seen fluctuations in recent years, a pattern that is likely to intensify as coffee crops globally experience disruption due to climate change. The third most significant crop is probably bananas.

There is limited animal husbandry, mainly chickens and pigs. A small minority engages in other occupations, most notably taxi driving, weaving and teaching.

Some Tenejapans have migrated to San Cristóbal and found work there in the service or construction industry while others engage in seasonal migration beyond Chiapas to work in a variety of manual labors. However this labor migration is much smaller than what has occurred in other highland municipalities such as Zinacatán.

The basic practice of subsistence farming is much the same as it was 40–50 years ago with a significant exception. Land shortages have meant that less land is put into fallow at any given time. In order to maintain productivity the Highland Maya have become increasingly reliant on chemical fertilizers (Collier 1994). In the short term this has been beneficial in increasing yields and providing food for a increasingly larger population. The long-term sustainability of this system is questionable and it has also made the Highland Maya subservient to a cash economy over which they have little control (Collier 1994). The implementation of the North American Free Trade Agreement in 1994 meant that imported maize grown in the United States with high fossil fuel inputs is cheaper than maize grown by less intensive methods within Mexico (Barry 1995). The impact of this in Highland Chiapas has been less severe than in other parts of Mexico simply due to the fact that most of the maize produced in the Highlands is for subsistence and local markets and not exported. The ownership of land in Tenejapa is divided between 70% as communal property, 20% as *ejido*, and 10% as private property. There are thousands of agricultural units in production and with a total of 99.4 km^2 in the municipality, tend to be small.

There is an increasing interconnection between Tenejapa and the larger regional, national and international contexts. While these interconnections are nothing new and a case could be made that they began with the Spanish invasion of the Highlands in the sixteenth century, the level of interconnectedness is certainly increasing. Collier (1975) did the first detailed study of this phenomenon with the neighboring Tzotzil and the process has continued unabated since then. For example, Guatemalan immigrants (mostly Mayan themselves although obviously from a different socio-linguistic group) are sometimes employed to pick coffee and for other agricultural labor in communities at lower elevations.

2.3.3 Ecology

This section presents some relevant biophysical, geographical, and ecological characteristics of Tenejapa; *ethno*-ecological classification is discussed in detail later in the book. The highland region of Chiapas presents a vast array of ecological zones and microenvironments and Tenejapa is no exception. Elevations range from 900 to 2500 m, with a few peaks reaching 2800 m. The municipality is almost equally divided in area between the three major ecological zones based on altitudinal and climatic variation: hot or *k'ixin k'inal* (900–1500 m), temperate (1500–1800 m) and cold or *sikil k'inal* (1800–2800 m). There is a general trend towards a downward slope moving from the south to the north of the municipality. Tenejapa lies along a seafloor uplift plateau of cretaceous limestone with a shallow soil profile punctuated by steep ravines and occasional karstic sinkholes, especially in the west central area (Fig. 2.13; Müllerried 1957). Soil is mainly nitosols with a high clay content which

Fig. 2.13 The patchy karstic landscape of Tenejapa as seen from a ridge in the community of Nabil facing southeast towards the primary forested peaks of Matzab

are susceptible to erosion and acrisols which are acidic and nutrient poor (Anon 1988). There is a pronounced rainy season from late May to December. Annual precipitation ranges from 1500 to 2000 mm and is higher in the northern, lower elevation parts of the municipality. The climate of Tenejapa is considered to be subhumid temperate, climatic formula $C(w_2)(w)$ in the Köppen climate classification as modified by García (1988). The central location of the municipality within the Highlands means there is flora represented from both the Selva Lacandon lowlands to the east and the Sierra Madre Mountains to the west. In general, the municipality is characterized by a patchy mosaic of land in cultivation and in various successional stages. Vegetational formations vary depending on the altitude and degree of human disturbance. In the lower elevations there is seasonal evergreen forest, tropical deciduous forest and pine-oak-liquidambar forest. At higher elevations lies pine-oak forest (although as oaks are selectively removed it is increasingly more common to find only pine forest, González-Espinosa et al. 1991) and evergreen cloud forest on the highest peaks where agriculture is impractical (Breedlove 1981). There are at least 2200 vascular plant species in Tenejapa (Berlin et al. 1974). Large animals have been hunted out of Tenejapa and it is rare to find even small mammals today, although Tzeltales will sometimes go to extraordinary lengths to hunt them despite the minimal return on their efforts (a potentially vexing issue for proponents of optimal foraging theory!). Deforestation is a serious problem as more and more land is brought under cultivation.

2.4 Research Communities in Tenejapa

Given the range of ecological variation present in Tenejapa it is necessary to conduct research in multiple communities in order to gain an accurate understanding of human-environment interactions. Much of this research was conducted mainly in six representative communities in Tenejapa (Fig. 2.14). Work was conducted in two

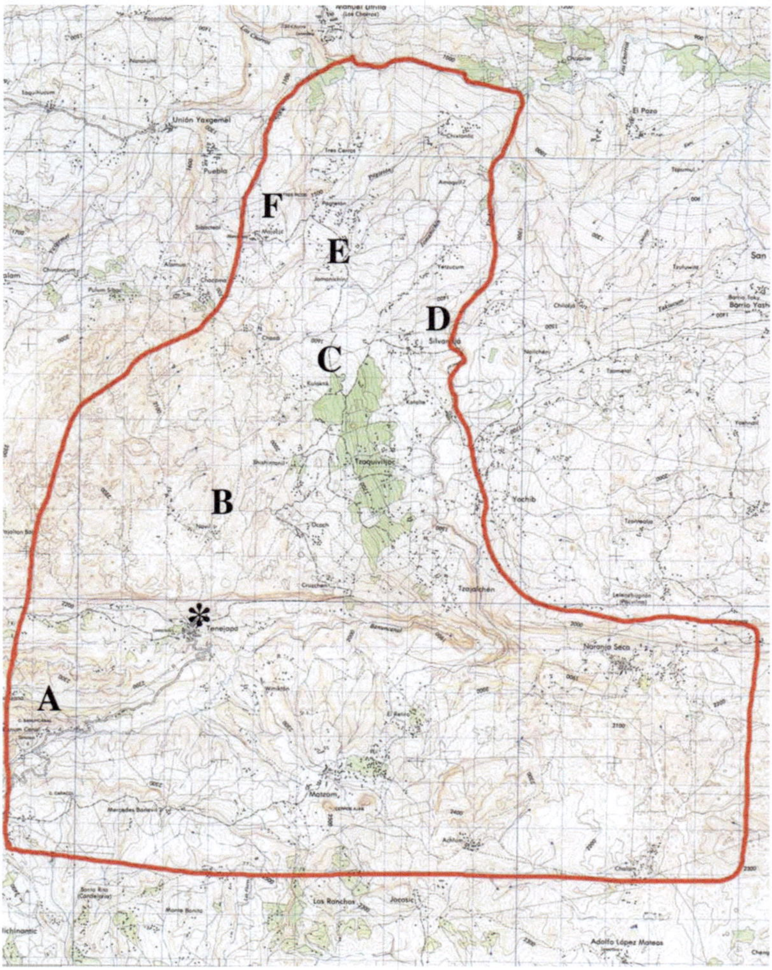

Fig. 2.14 Topographical map of approximate boundaries of Tenejapa with study communities: A—Balum Kanal, B—Nabil, C—Kul ak'tik, D—Sibinilja, E—Jomanichim, F—Majosik', * Tenejapa Center

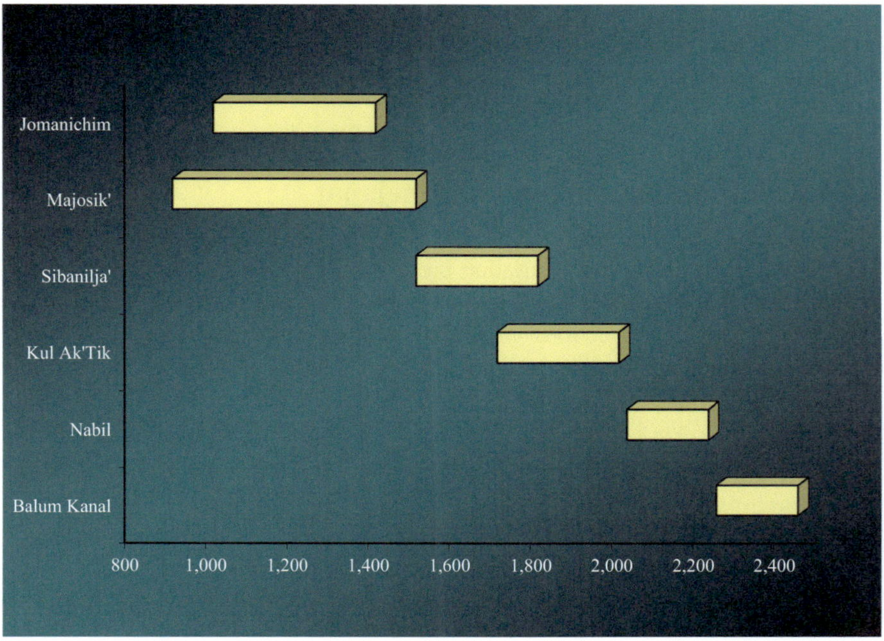

Fig. 2.15 Elevation range of study communities (in meters)

communities in each major ecological zone (six total) along a range of altitude were (Fig. 2.15). They are Balum Kanal and Nabil (cold country); Kul ak'tik and Sibanilja' (temperate country); Majosik' and Jomanichim (hot country).

*2.4.1 Cold Country (*Sikil K'inal*) Communities*

2.4.1.1 Balum Kanal

Balum Kanal (gloss 'nine levels', referring to the nine mythological levels that one ascents to the top of the mountain in this community although Becerra (1932) claims 'new houses up high') lies at 2240 m (16° 46′ 49″ North, 92° 32′ 12″ West) along the all weather paved road from San Cristóbal to Tenejapa center. This location makes it one of the most accessible communities in Tenejapa. Houses lie mostly along the road itself and the terrain is very steep. There is no clinic in Balum Kanal but there is access to the clinics in Tenejapa center. The population was 420 during the last census (INEGI 2011).

Fig. 2.16 Community of Nabil

2.4.1.2 Nabil

Nabil lies at an altitude of 2220 m (16° 50′ 17″ North, 92° 30′ 25″ West) along an all weather dirt road a few kilometers west of the main highway from Tenejapa to Oxchuc and just to the north from the box canyon where Tenejapa center lies (Fig. 2.16). Nabil contains some of the most karstic landscape in Tenejapa with numerous sinkholes. Despite the steep slopes of some of the sinkholes they are readily utilized for milpa. Most houses in Nabil have electricity and there is a clinic. The population was 473 during the last census (INEGI 2011).

2.4.2 *Temperate Country (*Sikil Htebuk K'inal/K'ixin Htebuk K'inal*) Communities*

2.4.2.1 Kul ak'tik

Kul ak'tik lies a short distance off the main paved highway from Tenejapa to Oxchuc at an altitude of 1800 m (16° 51′ 52″ North, 92° 29′ 12″ West; Fig. 2.17). The population was 844 during the last census (INEGI 2011). The community lies in the transition zone between cold and hot country and is the first area where coffee cultivation appears coming from the higher altitudes of the southern part of the municipality. There is no clinic.

Fig. 2.17 Community of Kul ak'tik

2.4.2.2 Sibanilja'

Sibanilja' lies at 1500 m (16° 52′ 28″ North, 92° 27′ 40″ West) close to the border with Oxchuc. There is a clinic available that is shared with the community of Pocolum (Fig. 2.18). The population combined with Pocolum was 1419 during the last census (INEGI 2011).

*2.4.3 Hot Country (*K'ixin K'inal*) Communities*

2.4.3.1 Jomanichim

Jomanichim (gloss 'place of the flower') lies at the intersection of an unpaved road off the main highway and a turnoff for the far northern part of the municipality at an altitude of 1420 m (the center lies at 16° 53′ 12″ North, 92° 28′ 41″ West). At times the road is not accessible during the rainy season. The population was 1112 during the last census (INEGI 2011). The community has suffered major political violence in the past as a result of conflicts between the dominant PRI political party and the opposition PRD political party.

Fig. 2.18 Sibanilja'/Pocolum clinic

2.4.3.2 Majosik'

Majosik' (gloss 'where the swallow was hit', Becerra (1932) claims 'hiding place of five swallows') is the largest community where I have conducted research, comprising over 1116 people at the last census count (INEGI 2011). It is located in the northwestern section of the municipality (the center lies at 16° 53′ 48″ North, 92° 29′ 32″ West) and ranges from 1500 to 900 m along a mountain slope that ends at the Majosik' river which forms the border between Tenejapa and the Tzotzil municipality of Chenalho. Majosik' was the site of Brian Stross' dissertation research in 1968 and 1969 on childhood acquisition of botanical terminology (Stross 1970). A 30 year follow-up study found, surprisingly, that children are continuing to acquire botanical knowledge at the same rate and levels (Zarger and Stepp 2004). More recent changes in Majosik' due to the growth of the coffee industry have led to outsourcing of labor to other non-Highland Mayans. Being in hot country, a range of agriculture is available, although coffee and maize predominate. Although Stross (1970) found that at the time of his research there was only one maize harvest a year, more recent years have seen two harvests throughout the year. It is not uncommon to see houses with large trucks parked in front, attesting to the greater economic opportunities afforded by hot country agriculture. Most houses have cement courtyards that provide an area for drying coffee beans and sometimes other crops as well (Fig. 2.19). Majosik' in some respects serves as a de facto *cabecera* for hot country Tenejapa. A variety of market and government services are available in Majosik', including a library, a guesthouse and a clinic.

Fig. 2.19 Courtyard in Majosik'

References

Adams C (2004) The ethnophysiology of the Tzeltal Maya of Highland Chiapas. PhD dissertation, University of Georgia

Anon (1988) Los municipios de Chiapas. Secretaría de Gobernación y Gobierno del Estado de Chiapas, Tuxtla Guitierrez

Barthlott W, Lauer W, Placke A (1996) Global distribution of species diversity in vascular plants: towards a world map of phytodiversity. Erdkunde 50:317–327

Barry T (1995) Zapata's revenge: free trade and the farm crisis in Mexico. South End Press, Boston

Becerra M (1932) Nombres Geográficos Indigenas del Estado de Chiapas. Gobierno del Estado, Tuxtla Guitierrez

Berlin B (1962) The Tenejapa dialect of Tzeltal: an outline of morphology (anthropology research projects). Stanford University, Stanford

Berlin B, Breedlove DE, Raven P (1974) Principles of Tzeltal plant classification. An introduction to the botanical ethnography of a Mayan-speaking people of the highlands of Chiapas. Academic, New York

Berlin EA, Berlin B (1996) Medical ethnobiology of the Highland Maya of Chiapas, Mexico: the gastrointestinal diseases. Princeton University Press, Princeton

Branstetter K (1974) Tenejapans on clothing and vice-versa; the social significance of clothing in a Mayan community. PhD dissertation, University of California, Berkeley

Breedlove DE (1981) Introduction to the flora of Chiapas. In: Breedlove DE (ed) Flora of Chiapas, part 1. California Academy of Sciences, San Francisco

Breedlove DE (1986) Listados Florísticos de México. IV. Flora de Chiapas. Instituto de Biología, UNAM, México

Brown P (1979) Language, interaction, and sex roles in a Mayan community: a study of politeness and the position of women. PhD dissertation, University of California, Berkeley

Brown P (2007) Principles of person reference in Tzeltal conversation. In: Person reference in interaction: linguistic, cultural, and social perspectives. Cambridge University Press, Cambridge, pp 172–202

Casagrande D (2002) Ecology, cognition, and cultural transmission of Tzeltal Maya medicinal plant knowledge. PhD dissertation, University of Georgia
Castille DM (1996) Psychological affliction and mental illness among Maya Indians of Highland Chiapas, Mexico. PhD dissertation, University of California, Berkeley
Collier GA (1994) Basta! Land the Zapatista rebellion in Chiapas. Food First Books, Oakland
Campbell L & Kaufman T (1985) Mayan linguistics: Where are we now? Ann Rev of Anthropology 14(1):187–198
Collier GA (1975) Fields of the Tzotzil: The ecological bases of tradition in highland Chiapas. University of Texas Press, Austin
D'Alesandro R, Cabañas AAG (2017) La Práctica de la Milpa: el ch'ulel y el maíz como elementos articuladores de la cosmovisión sobre la naturaleza entre los tzeltales de Tenejapa en los Altos de Chiapas. Estudios de Cultura Maya 50:271–297
Doane M (2011) Orphans in the global system: Maya coffee producers in Chiapas, Mexico. Anthropol Now 3(2):17–27
Fabian J (1983) Time and the other: how anthropology makes its object. Columbia University Press, New York
García E (1988) Modificaciones al Sistema de Clasificación Climática de Köppen. UNAM, Mexico
González-Espinosa M et al (1991) Secondary succession in disturbed Pinus-Quercus forests in the highlands of Chiapas, Mexico. J Veg Sci 2:351–360
Haehl JH (1980) A formal analysis of Highland Maya Kinship: Tenejapa as a special case. PhD dissertation, University of California, Irvine
Harlan J (1975) Crops and man. American Society of Agronomy, Madison
Hunn E (1977) Tzeltal Folk zoology. Academic, New York
INEGI (1993) La Población Hablante de Lengua Indígena en México. Instituto Nacional Estadística, Geográfica e Informática, Mexico
INEGI (2011) Mexico population and housing census 2010. Instituto Nacional Estadística, Geográfica e Informática, Mexico
INEGI (2015) Catálogo De Claves de Entidades Federativas, Municipios y Localidades. Instituto Nacional Estadística, Geográfica e Informática, Mexico
INEGI (2017) National survey of occupation and employment. Instituto Nacional Estadística, Geográfica e Informática, Mexico
Köhler U (2000) The Maya of Chiapas since 1965. In: Monaghan JD (ed) Supplement to the handbook of middle American Indians, vol 6. University of Texas Press, Austin, pp 179–206
Kraft KH, Brown CH, Nabhan GP, Luedeling E, Ruiz JDJL, d'Eeckenbrugge GC, Hijmans RJ, Gepts P (2014) Multiple lines of evidence for the origin of domesticated chili pepper, *Capsicum annum*, in Mexico. Proc Natl Acad Sci 111:6165–6170
Kuchka HE (2001) Method for theory: a prelude to human ecosystems. J Ecol Anthropol 5:1–78
Lampman A (2004) Tzeltal ethnomycology: naming, classification and use of mushrooms in the highlands of Chiapas. PhD dissertation, University of Georgia
Luber GE (2002) The biocultural epidemiology of 'second-hair' illness in two Mesoamerican societies. PhD dissertation, University of Georgia
Law D (2013) Mayan historical linguistics in a new age. Language and Linguistics Compass 7(3):141–156
Maffi L (1994) A linguistic analysis of Tzeltal Maya ethnosymptomatology. Anthropology. PhD dissertation, University of California, Berekeley
Medina A (1991) Tenejapa: Familia y Tradición en un Pueblo Tzeltal. Gobierno del Estado de Chiapas, Tuxtla Gutiérrez
Mittermeier RA (1988) Primate diversity and the tropical Forest: case studies from Brazil and Madagascar and the importance of the megadiversity countries. In: Wilson EO, Peter FM (eds) Biodiversity. National Academy Press, Washington, DC, pp 145–154
Müllerried FK (1957) Geología de Chiapas. Editorial Cultura, Mexico
Odum EP (1993) Ecology and our endangered. Life support systems, 2nd edn. Sinauer Associates, Sunderland

Page-Pliego JT (2013) Refresco y Diabetes entre los Mayas de Tenejapa, San Cristóbal de Las Casas y Chamula, Chiapas. Limina R 11(1):118–133
Patten BC (1978) Systems approach to the concept of environment. Ohio J Sci 78:206–222
Ramamoorthy TP, Bye R, Lot A, Fa J (eds) (1993) Biological diversity of Mexico: origins and distribution. Oxford University Press, New York
Rostas S (1986) From Ethos to Identity: religious practice as resistance to change in a Tzeltal community, Tenejapa, Chiapas, Mexico. PhD dissertation, University of Sussex
Ryan AT (1993) Supporting indigenous peoples. State of the world 1993. Worldwatch Institute, Washington, DC
Rzedowski J (1991) Diversidad y Orígenes de la Flora Fanerogámica de México. Acta Botánica Mexicana (14):3–21
Rzedowski J (1993) Diversity and origins of the phanoerogamic flora of Mexico. In: Ramamoorthy TP, Bye R, Lot A, Fa J (eds) Biological diversity of Mexico: origins and distribution. Oxford University Press, New York
Schizas DG, Stamou GP (2017) Network rethinking of nature and society. Ludus Vitalis 13(24):55–82
Simons GF, Fenning CD (eds) (2018) Ethnologue: languages of the world, twenty-first edition. SIL International, Dallas
Sisk TD et al (1994) Identifying extinction threats. Bioscience 44:592–603
Starr F (1902) Notes upon the ethnography of southern Mexico, vols 8 and 9, part 2. Davenport Academy of Sciences
Stepp JR (1999) Prospectus for information ecology. J Ecol Anthropol 3:39–74
Stepp JR (2016) Ethnobiological memoirs and memories. Ethnobiol Lett 7(2). art1
Stross B (1970) Aspects of language Acquisition in Tzeltal Children, PhD dissertation, University of California, Berkeley
Tax S (1937) Culture and civilization in Guatemalan societies. Sci Mon 48:463–467
Toledo VM (1988) La Diversidad Biológica de México. Ciencia y desarrollo 81(14):17–30
Toledo VM (1994) La Diversidad Biológica de México: nuevos retos para la investigación de los noventa. *Ciencias* 034
Yamakake K, Sánchez M, Ovando M, Hernández S, Antoniocoaut J (2009) Origen y Diversificación del Maíz: una revisión analítica. UNAM, Mexico
Zarger RK, Stepp JR (2004) Persistence of botanical knowledge among Tzeltal Maya children. Curr Anthropol 45:413–418

Chapter 3
Medicinal Plants and Preferred Habitats

A discussion of medicinal plant use should also involve an exploration of illness and the environment. In many ways, the environment in Highland Chiapas is marginal due to the relatively impoverished soils and cold climate in many parts. It is not surprising that the area was less inhabited prior to European conquest, with more people living at lower elevations where the soil quality is potentially higher and the weather warmer (Sharer 2006). The circumscription of native peoples in the Highlands by the Spanish while large landholdings were concentrated in the surrounding lowlands has implications to this very day. Many municipalities (although not all) have a high population density and little infrastructure in the way of potable water and sewage treatment. The results are environmental and social conditions that allow for high rates of illness. It is not uncommon to visit a household and find that someone has experienced some form of illness in the previous week. During one intensive study period of illness over several months, on average, households were free of illness for only approximately 15% of the time. This was a welcome event and allowed for interviews regarding why everyone in the household had remained healthy since my previous visit. This was usually attributed to good fortune and people did not articulate reasons beyond this.

The types of illnesses encountered in Tenejapa are familiar to anyone who has worked in marginalized montane tropical communities. The most common illnesses are gastrointestinal diseases (including general diarrhea, bloody and mucoid diarrhea, abdominal pain, epigastric pain, and intestinal parasites) and respiratory diseases (coughs, chronic cough such as croup, tuberculosis, inflammations of the nose, ears, and throat) (cf. Berlin and Berlin 1996). These two categories accounted for 87% of illnesses. Fever, eye infections, and mouth infections accounted for another 10%. The remaining 3% of illnesses involved skin infections, headaches, breaks and sprains, injuries from accidents and violence, teeth and mouth infections, mental illness, snakebites and illnesses with supernatural etiologies. Despite an inordinate amount of attention paid by researchers in Chiapas (e.g. Fabrega and Silver 1973; Holland 1963) to the treatment of illnesses with supernatural etiologies, Tzeltal Maya rarely experience such illnesses and attribute natural causes to

J. R. Stepp, *Ethnoecology and Medicinal Plants of the Highland Maya*, Ethnobiology, https://doi.org/10.1007/978-3-319-69315-6_3

most sickness. It is possible that supernatural illnesses are underreported due to the sensitive nature since witchcraft and evil spirits are the usual causes. However, it is very clear that the vast majority of illnesses in Tenejapa are naturalistic in etiology (cf. Foster 1976), empirically diagnosed and, at least initially, self-treated. Visits to specialized healers are not very common and are usually done specifically to treat illnesses with personalistic etiologies that would not likely be cured with medicinal plants. Western medicine clinics of varying quality exist throughout the Highlands and most Tzeltal can reach one within less than a few hours journey. However, despite fairly widespread access to clinics, Tzeltal rarely utilize them for everyday illnesses, preferring instead to self medicate with plants. Clinic visits are more common when physical trauma such as breaks and sprains are involved but, even then, there are a number of medicinal plants readily available for treatment.

Illness in Tenejapa does not occur with equal distribution throughout the year but is influenced by the shift in seasons. Illness increased during the month of July and peaked in August (Fig. 3.1). This coincides with the start of the tropical rainy season at the end of June/beginning of July. One likely explanation is that there is increased contamination of the water supply at the onset of the rainy season. Sewage and septic systems are virtually non-existent in Tenejapa outside of Tenejapa center. The most common place to defecate is in the milpa fields. During the dry season fecal matter stays relatively isolated in the milpa but with the onset of the rains it is flushed into the water supply and across the landscape. People may become exposed to gastrointestinal pathogens either by drinking contaminated water or by walking in contaminated soils and being exposed to mud mixed with feces. One of the results of environmental disturbance is the increase of water running across the landscape. In cases where the topsoil is eroded water is less likely to soak into the ground but instead runs over hard clay and limestone surfaces.

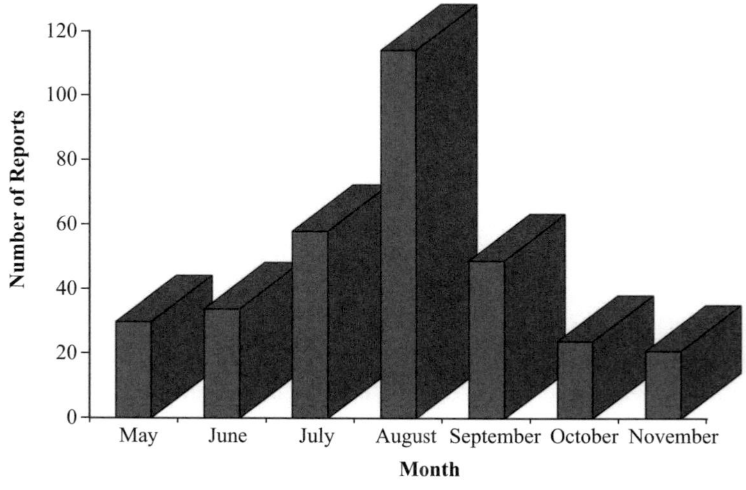

Fig. 3.1 Illness in Tenejapa

An increase in respiratory illnesses may also be related to the onset of the rainy season. Especially at higher elevations, the rainy season is marked by cloudy days, cold nights and perpetually damp surroundings. Apart from the general tendency of these environmental conditions to foster respiratory illnesses, people's daily household activities may be contributing as well. Most house compounds have a kitchen area where a lot of time is spent around an open-hearth fire with poor ventilation. During cold and rainy periods, people spend more time close to the fire and are exposed to more smoke.

3.1 Use Recall Interviews

A total of 208 individuals participated in use recall interviews, which are described in detail below. Participants were provided a rudimentary plant press and asked to collect and press a voucher specimen of every plant used to treat illnesses that occurred in the period between my visits. They were also told that if no illnesses occurred in the household in a given period then that was of interest as well. They were also asked to make a mental note about various ecological aspects about where the plant was obtained (habitat, soil type, abundance of plant) and the amount of time required to find the plant. Subsequent visits to the household involved a semi-structured interview process where informants were asked about all of the illness they treated in the week prior and the plants used to treat them.

A variety of use, management, selection, time allocation and ecological data were provided through these interviews. The following questions were asked in Tzeltal but I provide below the English translations as well (Stepp 2009):

1. *¿sbil chamel?*
   What is the name of the illness?
2. *¿sbil te mach'a ay yich'o chamel?*
   What is the name of the person with the illness?
3. *¿binti poxilil la sle?*
   How was the illness treated? While this almost always involved medicinal plants, occasionally they would report the use of patent medicines or injections.
4. *¿binti jejchuk la stuntes be te wamale?*
   If medicinal plants were utilized, what part of the plant was used?
5. *¿te' bal, wamal bal, ak' bal, binti?*
   Which of the three major plant life form categories the plant was in (*te'*–tree or woody shrub, *wamal*–herbaceous plant, *ak'*–grass)?
6. *¿yip?*
   What was the strength of the plant utilized?
7. *¿ay bal yorail te bit'il k'alal ay yip te wamale?*
   Is there a particular hour or season during which the plant was more effective?

8. *¿ay bal yawil ya xch'i te wamal te k'alal ay yipe?*
   Is there a particular place to collect the plant where the plant would be more effective?
9. *¿te 'nax bal la yich' tael ta comunidad te wamale?*
   Did you find the plant in the same community that you live in? If not, where?
10. *¿jayeb ora ya xbeben ta sleel te wamal te banti ay snae?*
    How long did it take you to find this plant walking from your house?
11. *¿ay bal yorail te k'alal ya yich' leele?*
    Are there special conditions to gather or harvest this plant?
12. *¿banti ya x'chi? (ak'beya yejtal te banti ya xch'i te wamale)*
    k'altik___ k'ajbenal____ wank'altik___ unin k'inal____ k'inal_____ te'tikil____tojol k'inal____ ja'mal_______pat na_____ kajpetal_____ akil_____ ts'unbil____otro ____________
    What habitat did you find the plant in?
13. *¿binti sbil lumilal banti ya xch'i?*
    What type of soil did it grow in?
14. *jich yipal (+++++ ya sk'an yal lom toyol, + ya sk'an yal lom alan)*
    How abundant is the plant (on a scale of 1–5)?

The voucher specimen collected by household members serves as a prompt for the interview and the use of such visual stimuli follows protocols set out in Whyte (1984), Boster (1987), Johnson and Griffith (1998), and Berlin and Berlin (1996). The protocol established by Berlin and Berlin is of particular importance for this study. They developed what they referred to as an *hebario viajero* or 'traveling herbarium' that consisted of dried herbarium vouchers with the labels removed of the 203 most common medicinal plants as determined in their long-term studies on Highland Maya medical ethnobotany (Berlin and Berlin 1996:81–2). This enabled them to collect a wide range of ethnomedical and ethnoliguistic data throughout the Highlands. The success of this methodology influenced me to develop a similar protocol. The voucher specimen collected by informants, while often sterile, was usually easy to determine. Because my collaborators always have a good working knowledge of the common flora of Tenejapa, of which a substantial subset is the medicinal flora, and my own familiarity with the most utilized medicinal flora, we are usually able to provide botanical determinations of plants that are similar in appearance when sterile. On the rare occasion when we are unable to make a determination based on a voucher collected by a participant, we seek out the actual location where the plant was collected and try to make another collection for determination. If this still proves unsuccessful we note the plant population with GIS coordinates and return to collect a fertile specimen. While collection of sterile specimens is usually to be avoided, it is necessary in studies like these. Moreover, previous research in Tenejapa and the Highlands has resulted in over 7000 collections with a reference collection readily available at the El Colegio de la Frontera Sur (ECOSUR) herbarium to cross check determinations. Contextualized interviews are also often done at irregular intervals and involve traveling to the particular ethnoecological zone where the plant is found and discussing salient features of the zone with the informant. This allows for a control on informant's reports of their behavior versus their actual behavior.

3.2 Procurement and Use of Medicinal Plants

For all other cases of illnesses besides the ones that involve clinic visits or visits to specialized healers, people self treated themselves with medicinal plants. In all, we recorded 289 discrete illness cases that were treated with medicinal plants. Three hundred and twenty-seven plants were used in treatments, comprising 122 species. The species that were identified and accessioned at ECOSUR are shown in Table 3.1, which provides the scientific name; the three-letter abbreviation for the plant family based on Weber (1982); the Tenejapa Tzeltal common name; and the number of times the plant was utilized. In 63% of the treatments only one medicinal plant was used. In 32% of the treatments an admixture was made of a medicinal plant with at least one other medicinal plant.

In 5% of the treatments three medicinal plants were used together. More than three medicinal plants were never used in an admixture during the study. However, Tzeltal Maya healers who specialize in herbal medicine often add more than three plants together. This is not surprising given the need to create added value when the general population is already familiar with many medicinal plant cures. In 4% of the treatments an admixture was made of one medicinal plant with a non-plant substance. The most common is salt, although we have found that horse feces and sugarcane liquor (*pox*) are also used idiosyncratically.

3.2.1 Parts of Plants Utilized

The Tzeltal Maya do not generally use dried medicinal plant material but instead gather fresh material as needed. This befits medicinal plants as belonging to a commons where any community member can freely access them. Some 'entrepreneurial' healers have tried to repackage dried plant material into gelatin capsules and other delivery systems but these attempts at creating at market economy of freely available medicinal plants have largely failed and for good reason. Occasionally, a sale is made to ignorant tourists which may be just enough to keep certain individuals motivated to keep trying.

The Tzeltal have an in-depth understanding of plant morphology and recognize at least 116 distinct plant parts (Berlin et al. 1974). The most important plant parts utilized for medicinal purposes are *y'abenal* (leaves), accounting for 74% of uses. This is followed by *y'isim* (roots) which account for 5% of uses. The remaining 20% is accounted by the following parts in order of importance: *sni'* (tip of plant leaves or flowers), *sit* (fruit), *ste'el* (stems), *ye'tal* (bulb or tubers), *spat* (tough non-leathery bark), *snich* (flowers), *sbak'* (seeds), *ya'lel* (juice of fruit).

Table 3.1 Medicinal plants utilized in Tenejapa

Scientific name	Fam.	Tenejapa Tzeltal name	Times used
Abelmoschus moschatus Medik.	MLV	kaxlan tunim	5
Ageratina ligustrina (DC.)	AST	cha' te'	1
Allium sativum L.	ALL	axux	1
Aloe vera (L.) Burm. f.	LIL	sabila	3
Ambrosia cumanensis Kunth	AST	altamixa	5
Apium leptophyllum (DC.) F. Muell. Arg. ex Benth.	API	kulantu chitam	1
Artemisia absinthium L.	AST	ajeno	6
Baccharis confertoides Nesom	AST	mes te'	13
Baccharis trinervis (Lam.) Pers.	AST	sakil xijch wamal	4
Begonia heracleifolia Schltdl. & Cham.	BEG	majben wamal	2
Bidens pilosa L.	AST	majtas	1
Boehmeria ulmifolia Wedd.	URT	poxil tsa'nel	1
Borreria laevis (Lam.) Griseb.	RUB	buluk' sit wamal	3
Botrychium decompositum Mart. & Gol.	OPH	tsajal akan tsib	2
Bougainvillea glabra Choisy	NYC	bugambilla	2
Bouvardia leiantha Benth.	RUB	tsajal nich wamal	4
Brugmansia candida Pers.	SOL	kampana nichim	3
Bryophyllum pinnatum (Lam) Kurz	CRA	poxil majben	1
Buddleia crotonoides A. Gray	LOG	bak te'	2
Bunchosia lanceolata Turcz.	MLP	sak ji'te	1
Byrsonima crassifolia (L.) H.B.K.	MLP	lancin	3
Canavalia ensiformis (L.) DC.	FAB	sakil chenek'	2
Castilleja arvensis C. & S.	SCR	tzajal nich wamal	1
Cestrum noctrum L.	SOL	ch'opak te'	1
Chenopodium ambrosioides L.	CHN	kajk'an	1
Cissampelos pareira L.	MEN	chin ak'	9
Citrus limetta Risso	RUT	elomonex	3
Citrus limonia Osbeck	RUT	lima	2
Citrus sinensis (L.) Osbeck	RUT	alchax	3
Clethra suaveolens Turcz.	CLE	k'ak'et te'	1
Cuphea aequipetala Cav.	LYT	tsajal nich wamal	6
Cupressus benthamii Endl.	CUP	nukujpat	5
Cymbopogon citratus (DC.) Stapf	POA	limón ak	3
Drimys granadensis L. f.	WIN	tsajal on te'	1
Equisetum hyemale L.	EQU	tujt	1
Equisetum myriochaetum S. & C.	EQU	tujt	2
Erechtites hieraciifolius (L.) Raf. ex DC.	AST	cha'lam tsots wamal	1
Erigeron karwinskianus (DC.)	AST	sakil nich wamal	4
Eucalyptus globulus Labill.	MRT	kampor	4
Euphorbia hypericifolia L.	EUP	antun wamal	2

(continued)

Table 3.1 (continued)

Scientific name	Fam.	Tenejapa Tzeltal name	Times used
Fleischmanniopsis leucocephala (Benth.) King & H. Rob.	AST	sakil chixal nich wamal	1
Foeniculum vulgare L.	API	inojo	10
Gnaphalium purpureum L.	AST	sakil wamal	2
Hamelia patens Jacq.	RUB	tsajal nich wamal	1
Hedychium coronarium Koenig	ZIN	ja'ben	5
Iostephane trilobata Hernsley	AST	poxil me'winik wamal	1
Kearnemalvastrum lacteum (Aiton) Bates	MLV	mal wax wamal	1
Lagascea helianthifolia H.B.K.	AST	jos wamal	1
Lantana camara L.	VRB	ch'il wet	8
Lepidium virginicum L.	BRA	poxil pumel	1
Liquidambar styraciflua L.	HAM	sok te'	1
Litsea glaucescens H.B.K.	LAU	tsiltsil ujch	2
Mangifera indica L.	ANA	manko	1
Matricaria courrantiana DC.	AST	manzanilla	1
Mentha citrata Ehrh.	LAM	wena	5
Mentha spicata L.	LAM	wena	2
Monnina xalapensis H.B.K.	PLG	sts'ots' te'	1
Montanoa leucantha (Lag.) Blake	AST	xakrix te'	1
Musa paradisiaca L.	MUS	manzana lobal	1
Myrica cerifera L.	MYR	ch'aj kolol te'	2
Nama dichotomum (R. & P.) Choisy	HYD	sansira ak'	2
Nemastylis silvestris Loes.	IRI	pech'ech ak'	1
Nicotiana tabacum L.	SOL	bankilal	3
Ocimum selloi Benth.	LAM	san miguel wamal	1
Oenothera rosea L'Her. ex Aiton	ONA	tsajal nich wamal	6
Opuntia sp.	CAC	pejtak	1
Passiflora edulis Sims.	PAS	karanato	1
Persea americana Mill.	LAU	bak' on	1
Petiveria alliacea L.	PHT	pay te'	1
Phenax mexicanus Wedd.	URT	tsots wamal	1
Phyllanthus lathyroides H.B.K.	EUP	poxil xiwel wamal	1
Pinaropappus spathulatus var. *chiapensis* McVaugh	AST	tsis kojtom wamal	1
Plantago major L.	PLA	poxil bejts'em	1
Pluchea odorata (L.) Cass.	AST	sitit	1
Polygala costaricencis Chodat	PGL	poxil ea'l	1
Polygala floribunda Benth.	PGL	ch'opak te'	2
Polygala paniculata L.	PGL	poxil eal wamal	1
Priva aspera H.B.K.	VRB	jolom ik' wamal	1
Prunus persica (L.) Batsch	ROS	turesna	7
Prunus serotina Ehrh.	ROS	chichoj te'	1

(continued)

Table 3.1 (continued)

Scientific name	Fam.	Tenejapa Tzeltal name	Times used
Psidium guajava L.	MRT	pataj	2
Psidium guineense Sw.	MRT	pajchak'	5
Rapanea juergensenii Mez	MRS	xyax te'	1
Ricinus communis L.	EUP	ch'opak te'	1
Rumex crispus L.	PLG	yak' tsi' wamal	1
Rumex obtusifolius L.	PLG	yak' ts'i' wamal	1
Salmea scandens (L.) DC.	AST	ijk'al wamal	2
Salvia cacaliaefolia Benth.	LAM	jolom ik' wamal	1
Salvia cinnabarina M. & G.	LAM	sabal ts'unun	1
Salvia lavanduloides H.B.K.	LAM	ch'abakal	15
Sambucus mexicana C. Presl ex DC.	CAP	chijil te'	1
Sanicula liberta C. & S.	API	yax wamal	1
Smallanthus maculatus (Cav.) H. Robinson	AST	balan k'in	6
Solanum americanum Miller	SOL	moen	1
Solanum lanceifolium Jacq.	SOL	tujkulum ch'ix	4
Stevia ovata Willd.	AST	ch'aal wamal	1
Struthanthus quercicola (C. & S.) Naud.	LOR	yijkats te'	1
Symplocos breedlovei Lundell	SYM	pay te'	6
Taraxacum officinale Webb.	AST	lantsin pox	1
Tithonia diversifolia (Hemsl.) A. Gray	AST	ch'ajkil	12
Trichilia havanensis Jacq.	MEL	xenel pox wamal	1
Valeriana scandens L.	VAL	alambre ch'ox	1
Verbena carolina L.	VRB	sakil yakan k'ulub wamal	6
Verbena litoralis H.B.K.	VRB	yakan k'ulub wamal	33
Vernonia leiocarpa DC.	AST	bak te' wamal	1
Vernonia patens H.B.K.	AST	sitit	1
Viburnum hartwegii Benth.	CPR	tsop te'	1
Vitis bourgaeana Planchon	VIT	ts'usub	1
Zanthosylum foliolosum J.D. Smith	RUB	elomonex	1
Zea mays L.	POA	ixim	1

3.2.2 Medicinal Plants and Life Forms

The Tzeltal Maya recognize four major life form taxa: *te'* (trees and shrubs), *ak'* (vines), *ak* (grasses), *wamal* (herbaceous non-woody plants). The most significant category that utilized medicinal plants are placed in is *wamal,* accounting for 63%. The next category is *te'* accounting for 25%. This category is subdivided into *ch'in te'* (small tree or shrub) 43% of the time. The last category is *ak'*, representing 4% of medicinal plants utilized. We have no reported use of plants in the *ak* (grass) taxon.

3.2.3 Reported Strength of Medicinal Plants

In order to describe the strength or efficacy of a medicinal plant, the term *yip* (lit. strength) is employed. A scale of one to five (one lowest, five highest) is used in interviews to elicit *yip*. The usefulness of this question frame could be debated. On one hand, medicinal plants are always considered to have some amount of *yip*, or they would not be used. Also, the translation of the term is context dependent. While it appears that *yip* does relate to efficacy, it is polysemous and can also refer to a strong taste or pungency. Thus, a plant could be relatively ineffective in treating an illness but by virtue of its strong effects on the senses, still be considered to have high *yip*. The average *yip* of the medicinal plants was 4.18 for plants used as primary treatments; 4.32 for plants used as secondary treatments in admixtures; and 3.90 for plants used as tertiary treatments in admixtures. There does not appear to be a pattern regarding a particular species as high or low *yip*. All of the plants that receive a low (one or two) *yip* score by a particular person were also considered to be high *yip* (four or five) by at least one other person on at least one other occasion.

3.2.4 Medicinal Plants and Soil Types

The Highland Maya conceptualize a detailed soil taxonomy that is based on color, texture and organic content (Cervantes Trejo 1997) and employ it in agricultural activities, but it appears to be little utilized in the procurement of medicinal plants. The overwhelming response when asked about what kind of soil a particular medicinal plant grows in is *ijk'al lumilal* (black soil). Only twice has a participant replied otherwise, both times with *tsajal lumilal* (red soil).

3.2.5 Reported Abundance of Medicinal Plants

A scale of one to five (one lowest, five highest) is used in interviews to elicit the abundance of a plant when it is collected. Participants estimate the relative abundance of the plant resource in general, not just in the specific location where it is collected. The average abundance is 4.11 for plants used as primary treatments; 4.08 for plants used as secondary treatments in admixtures; and 3.89 for plants used as tertiary treatments in admixtures. This question frame is useful in identifying plants with low reported abundance. The following plants have a high degree of consensus as to their low (score of 1 or 2) relative abundance: *Abelmoschus moschatus*, *Ambrosia cumanensis*, *Botrychium decompositum*, *Hedychium coronarium*, *Castilleja arvensis Liquidambar styraciflua*.

3.2.6 Seasonality and Time of Collection

There is abundant literature in biochemical ecology noting that secondary defense compounds are not distributed evenly in a given species or in a given population (this is discussed in more detail in Chap. 6) but are subject to a wide range of fluctuation depending on genetic variability, location, resource availability, seasonality, and time of day (Harborne 1993). These last two factors are responsible for tremendous variation even in the same plant. For example, a study of the content and composition of alkaloids in hemlock (*Conium maculatum*) found that at certain times of the day alkaloids were non-existent, while at other times of the day their levels reached levels that would be highly toxic to a person who consumed the plant (Fairbairn and Suwal 1961). They also noted variation in composition and content of alkaloids based on seasonality (wet season versus dry season). This has major implications for medical ethnobotany if these findings are generalizable to other medicinal plants. Curious to find out if the Tzeltal had discovered that the bioactivity of medicinal plants fluctuated I have asked if, for a given plant, there is a time of day or night, or particular season when the plant is more or less effective at healing. No one has ever responded affirmatively to this question. This does not conclusively show that the Tzeltal do not identify fluctuations in plant biochemistry but it does suggest that if they do, it is for a limited subset of medicinal plants and that the knowledge is not widespread.

3.2.7 Time Allocation

For each medicinal plant treatment, participants were asked how long it took (in one direction) to gather the plant when walking from their household. This time data is then correlated with distance traveled by estimating the amount of time one can walk in an hour across the hilly terrain of Tenejapa. We do this by taking GPS measurements whenever we walk for a long distance with participants. On average, people walk 4 km an hour across the hilly terrain of Tenejapa, which, while not uniform, does consistently present similar elevational gradients when one takes an extended hike. Thus 15 min corresponds to 1 km, 30 min to 2 km, etc.

The average amount of time spent to find a specific plant for a treatment is 16 min (Fig. 3.2). A 1 km radius accounts for 71% of all medicinal plant procurement. Ninety-nine percent of medicinal plant procurement is conducted within a 2-km radius around each household On very rare occasions, people travel to other communities to obtain medicinal plants. The most time someone has spent procuring a medicinal plant is 6 h. This time was spent on obtaining *Equisetum myriochaetum*, a type of horsetail plant that only grows in marshy areas, an uncommon habitat in Tenejapa. Other instances where someone spent more than an hour procuring a plant involved travelling to areas at lower or higher elevations to find a particular plant. The ability to find significantly different environmental conditions due to

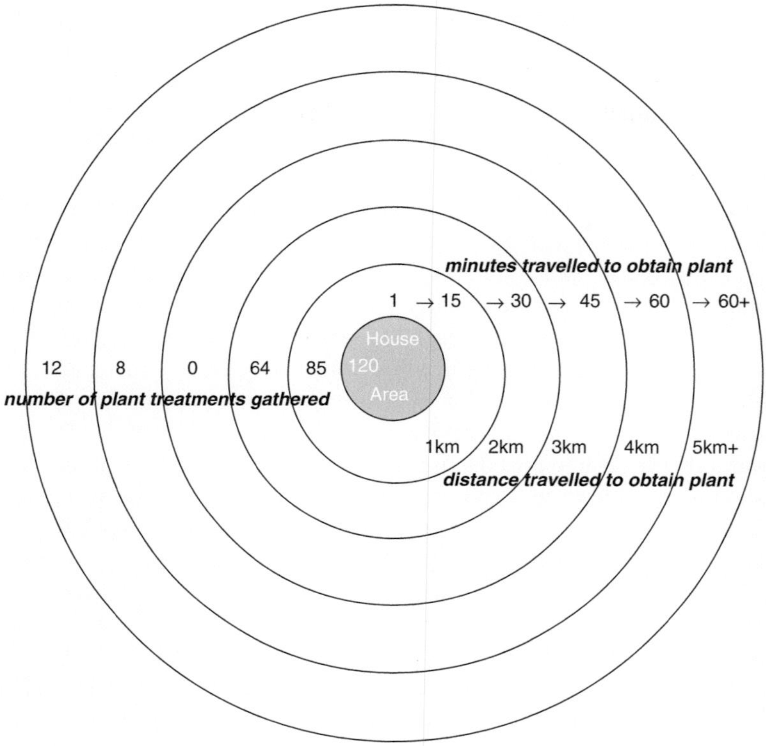

Fig. 3.2 Time and distance traveled to obtain medicinal plant treatments

elevation change in a relatively short straight line distance is a significant benefit for montane communities reliant on local flora for their survival.

It appears that the 'living pharmacy' of the Tzeltal lies within a 2 km vicinity of a given community. As noted above, medicinal plants generally have no market value to the Tzeltal and I have found no instances of medicinal plant purchases at markets, either in Tenejapa or at the regional markets of San Cristóbal, the de facto urban capital of the Highlands. The latter market however does contain a number of vendors selling dried medicinal herbs, their main customers being urban *ladinos* and tourists, not Highland Mayans.

3.2.8 Medicinal Plant Habitats

The Tzeltal Maya like many other indigenous and traditional groups, have a detailed understanding of ecological variation and this is exemplified in their ethnoecological classification system (see as examples Martin 1995; Hunn 1990; Meilleur 1986). Some studies have found that the number of named habitats by indigenous peoples can exceed that of the scientific literature (Shepard 2001; Fleck and Harder 2000).

Perceptions of the biophysical environment *vis a vis* botanical knowledge can be divided into five domains: soils, climates, vegetation types, land use and stages of ecological succession (Martin 1995). It should be noted that this typology does not necessarily correspond to emic ethnoecological classification but it a useful point of departure. I focus here on three domains: climate, stages of ecological succession and land use.

As noted in Chap. 2, the Tzeltal recognize three major ecological zones based on climate and elevation. These are hot country, *k'ixin k'inal* (900–1500 m); temperate country, called *k'ixin htebuk k'inal* or *sikil htebeuk k'inal*, glossed as a somewhat hot country or somewhat cold country (1500–1800 m); and cold country or *sikil k'inal* (1800–2800 m). Despite the lack of a refined term for the transitional temperate zone, the Tzeltal readily recognize this zone and the transitional nature of it.

Within each of these three major zones are a wide range of habitats. The Tzeltal are astute observers of vegetational succession and this is reflected in the lexicon. However, I do not wish to give the impression that the Tzeltal have a linear conceptualization of the landscape. Landscape successional processes are very much considered to be cyclical and are understood to be in constant flux (Fig. 3.3).

Allow me to further elaborate on this figure. I arbitrarily begin this discussion of the landscape cycle at the point at which fields are burned and prepared for milpa agriculture. The sub-cycle of burning, harvest and planting the *k'altik* (milpa) can continue up to 3 years but this time is lengthening as more and more Tzetal rely on chemical fertilizers that artificially boost the productivity of the milpa fields beyond what is provided through the release of nutrients during burning (Fig. 3.4).

After the land is put into a fallow a number of terms are used depending on the amount of time the land has been in fallow and the types of vegetation emerging.

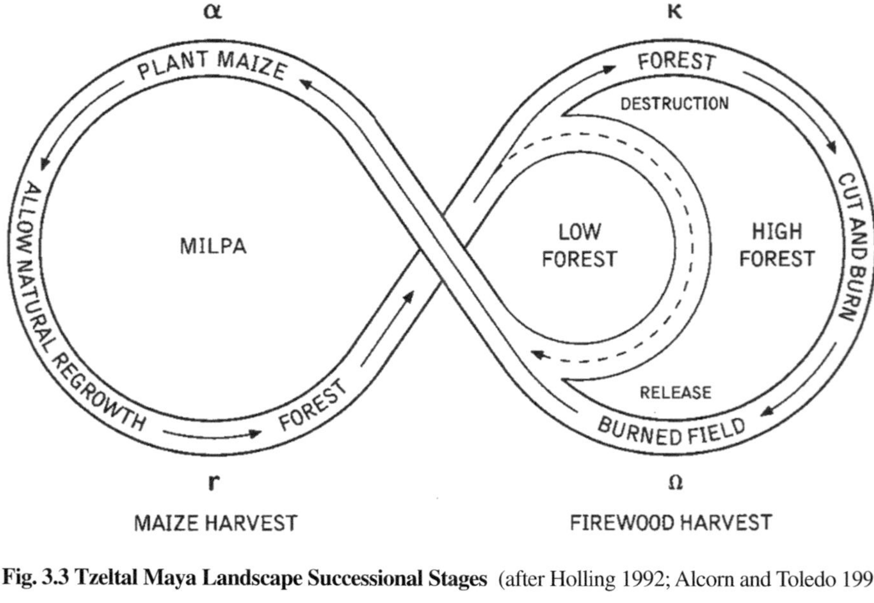

Fig. 3.3 Tzeltal Maya Landscape Successional Stages (after Holling 1992; Alcorn and Toledo 1998)

Fig. 3.4 First planting of a milpa after a burn

Fig. 3.5 First year fallow or *k'ajbenal*

The first year of fallow is called *k'ajbenal* (Fig. 3.5). This is followed by *wank'altik*, second year fallow. The next two categories overlap. They are *unin k'inal*, which is a 3–7 year period secondary growth vegetation and *k'inal* which is the 6–12 year period (Fig. 3.6).

Fig. 3.6 3–7 year (*unin k'inal*) and 6–12 year secondary growth (*k'inal*)

After this period when woody shrubs and small trees emerge the habitat is termed *te'tikil*, which glosses as young secondary forest (Fig. 3.7). Sometimes the modifier *ch'in* (small) is applied to particular young *te'tikil*. The structure and composition of *te'tikil* varies with altitude but there is no linguistic distinction made between hot country young secondary forest and cold country secondary forest. *Te'tikil* in hot country intermingles with areas of coffee production where *Inga* sp. is planted to shade the coffee plants (Fig. 3.8).

After a period of 30–50 years, te'tikil turns into *tojol k'inal* (lit. straight land), or old secondary forest (Fig. 3.9). Mature or primary forest that has never been cut or not been cut for many generations is called *ja'mal* (Fig. 3.10). This habitat is discussed in more detail in Chap. 5.

The Tzeltal recognize other habitats based on either land use or the dominant vegetation growing in it. These are: *kajpetal* or coffee grove; *pat na* which sometimes is used to refer to the home garden but generally refers to the immediate area around the household complex; *akil* or grassland (Fig. 3.11) and *ts'unbil*, meaning cultivated. This latter term does not actually refer to a habitat but allows us to make a distinction between wild volunteer plants growing in the *pat na* or vicinity of the home and plants that are intentionally cultivated (Fig. 3.12).

Fig. 3.7 Young secondary forest (*te'tikil)*

Fig. 3.8 Hot country young secondary forest (*te'tikil*)

Fig. 3.9 Old secondary forest (*tojol k'inal*)

Fig. 3.10 Mature forest (*ja'mal*)

Fig. 3.11 Grassland (*akil*)

3.2.9 Procurement of Medicinal Plants

Medicinal plants are obtained from a wide variety of habitats but there are very clear patterns present. If the various managed areas and early successional stages are aggregated together, this accounts for where 86% of medicinal plants are collected (Fig. 3.13). Only 14% of plants are collected in young secondary forest. There was only one collection of a medicinal plant in old secondary forest and no collection of medicinal plants in mature forest during the study.

If these categories are disaggregated even more patterns emerge (Fig. 3.14). The single most significant habitat in which to obtain medicinal plants is *pat na*, the immediate household area. This accounts for 35% of all collections. The next most important habitats were *k'inal*, where 21% of medicinal plants are collected, and *te'tikil*, where 14% are collected. *Akil* (grasslands) accounts for 10% of collections while another 10% of plants came from cultivation. The Tzeltal do not systematically practice medicinal plant cultivation but there are a few species that are regularly cultivated.

The cultivated species used as medicinal plants are *Abelmoschus moschatus, Allium sativum, Aloe vera, Ambrosia cumanensis, Artemisia absinthium, Brugmansia candida., Citrus limetta, Eucalyptus globulus, Foeniculum vulgare, Musa paradisiaca, Nemastylis silvestris, Nicotiana tabacum, Persea americana, Sambucus mexicana, Taraxacum officinale, Trichilia habanensis*. These plant fall into four main categories: old world introductions with a long history of medicinal use (*Allium sativum, Aloe vera, Artemisia absinthium, Eucalyptus globulus, Foeniculum vul-*

Fig. 3.12 Women and children planting medicinal plants in community of Balum Kanal, Tenejapa

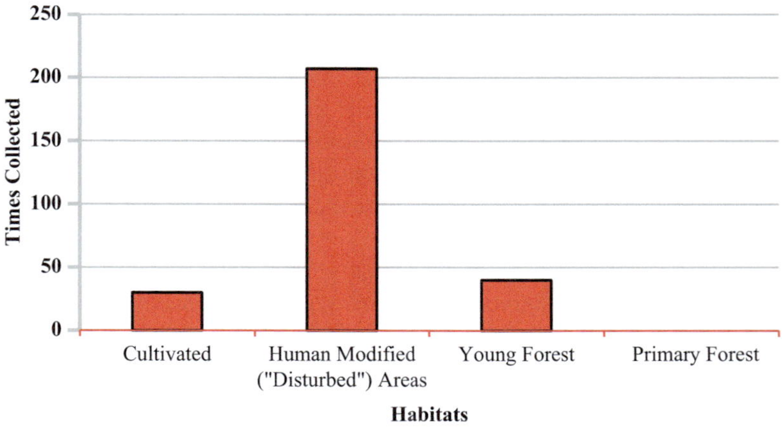

Fig. 3.13 Procurement of medicinal plants in aggregate habitat categories

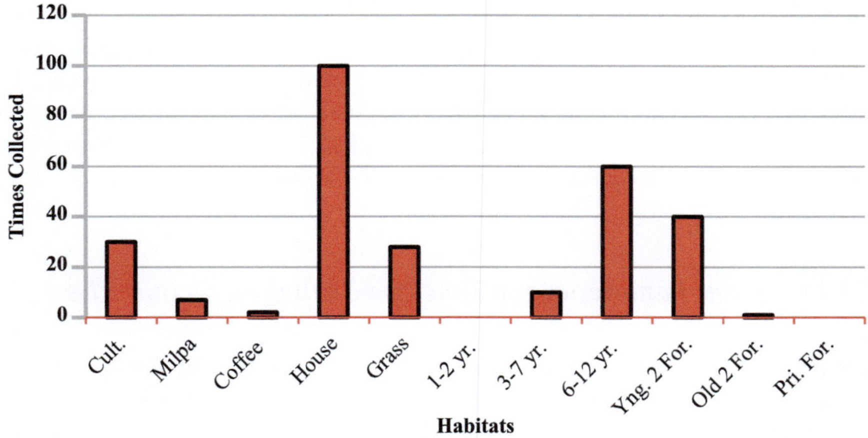

Fig. 3.14 Procurement of medicinal plants in disaggregated habitat categories

gare, Taraxacum officinale); introduced plants that are also utilized as foods (*Citrus limetta*, *Musa paradisiaca*), native plants that are also utilized as foods (*Persea americana*, *Sambucus mexicana*), and native plants that are used as medicinals that have been transplanted or brought into cultivation for ease of access (*Abelmoschus moschatus*, *Ambrosia cumanensis*, *Nemastylis silvestris*, *Nicotiana tabacum*, *Trichilia habanensis*).

3.2.10 Weeds and the Medicinal Flora

A striking feature of the Tzetal medicinal flora is the large number of plants that are herbaceous and grow in disturbed habitats. On way of analyzing the flora is looking at how many are considered to be weeds. Although weeds are a cultural construct, they have features in common that allow for definition. Baker (1965:147) defines a weed as "a plant…if, in any specified geographical area, its populations grow entirely or predominantly in situations markedly disturbed by man (without, of course, being a deliberately cultivated plant)." Weeds are those plants that are successful in disturbed environments, are fast growing, and, are often but not always herbaceous (Zimdahl 2018). I might add that a plant, just by virtue of being herbaceous and growing in a disturbed area, does not automatically qualify as a weed. Some degree of invasiveness is also required.

One hundred and nine different plant species that were regularly used were compared with a definitive checklist of weeds found in Mexico and Chiapas developed by Villaseñor and Espinosa (1998). A chi-square test was performed to determine the statistical significance of the distribution of weeds within the Highland Maya medicinal flora in relation to the distribution of weeds within the Chiapan flora as a whole (Stepp and Moerman 2001). Chiapas as a whole contains approximately

9000 vascular plant species (Breedlove 1981). Of these, 1178 (13%) are considered weeds (Villaseñor Ríos and Espinosa García 1998). One could predict then, that if weeds were randomly distributed in the Tzeltal medicinal flora, there would be about 14 weed species among these 109 medicinals (Table 3.2). In fact the actual number is 41, much higher than predicted ($p < 0.001$). Table 3.3 lists these plant species that are both weeds and medicinals for the Tzeltal Maya.

3.2.11 Plant Families of the Tenejapa Tzeltal Medicinal Flora

At the family taxonomic level, one can make some very broad and widespread generalizations about the characteristics of plants found within a given family. For example, the Asteraceae (or Compositae) is characterized primarily by herbaceous plants and evergreen shrubs with broad distribution. The Poaceae (or Gramineae) is characterized by grasses and is the most ecologically dominant group of plants in the world (Heywood 1993). Plant families have also been categorized successfully according to their importance for medicinals (Moerman 1991, 1996; Moerman et al. 1999). The most important families for medicinal plants in Tenejapa are presented in Fig. 3.15. This finding corresponds with an earlier study of the entire Highland Maya medicinal flora that also found that Asteraceae, Lamiaceae, and Solanaceae were the three respective top medicinal plant families (Moerman et al. 1999). Absent from the top rankings of the Tenejapa data is Rosaceae, which was ranked fourth in the Berlin and Berlin data used in the Moerman et al. (1999) paper. This difference can be accounted for by the fact that the Berlin and Berlin dataset is a comprehensive list of all plants *mentioned* as medicinals while the dataset here only contains medicinal plants that were *utilized* not just reported to be medicinal.

Many of the same top families for medicinal plants are the same top families for weeds (Holm 1978; Holm et al. 1979), with the exception of Poaceae. Among these families are Asteraceae, Fabaceae, Convolvulaceae, Euphorbiaceae, Chenopodiaceae, Malvacae, and Solanaceae (Stepp and Moerman 2001).

Table 3.2 Weed species in Tzeltal Maya medicinal flora (Stepp and Moerman 2001)

	Weed	Not weed	Total
Medicinal	41	69	110
Not medicinal	1143	7748	8897
Total	1178	7822	9000

$\mathbf{X}^2 = 56.70$; $p < 0.001$

Table 3.3 Weeds used as medicinal plants in Tenejapa

(Asteraceae) *Ambrosia cumanensis* Kunth
(Asteraceae) *Baccharis confertoides* Nesom
(Asteraceae) *Baccharis trinervis* (Lam.) Pers.
(Asteraceae) *Bidens pilosa* L.
(Crassulaceae) *Bryophyllum pinnatum* (Lam.) Kurz
(Scrophulariaceae) *Castilleja arvensis* C. et S.
(Chenopodiaceae) *Chenopodium ambrosioides* L.
(Menispermaceae) *Cissampelos pareira* L.
(Lythraceae) *Cuphea aequipetala* Cav.
(Asteraceae) *Erechtites hieracifolia* Raf.
(Asteraceae) *Erigeron karvinskianus* DC
(Umbelliferae) *Foeniculum vulgare* L.
(Rubiaceae) *Hamelia patens* Jacq.
(Malvaceae) *Kearnemalvastrum lacteum* (Aiton) Bates
(Verbenaceae) *Lantana camara* L.
(Brassicaceae) *Lepidium virginicum* L.
(Hamamelidaceae) *Liquidambar styraciflua* L.
(Lamiaceae) *Mentha spicata* L.
(Asteraceae) *Montanoa leucantha* (Lag.) Blake
(Myricaceae) *Myrica cerifera* L.
(Solanaceae) *Nicotiana tabacum* L.
(Lamiaceae) *Ocimum selloi* Benth.
(Onagraceae) *Oenothera rosea* L'Her. ex Aiton
(Phytolaccaceae) *Petiveria alliacea* L.
(Plantaginaceae) *Plantago major* L.
(Asteraceae) *Pluchea odorata* (L.) Cass.
(Polygalaceae) *Polygala paniculata* L.
(Rosaceae) *Prunus persica* (L.) Batsch
(Myrtaceae) *Psidium guajava* L.
(Euphorbiaceae) *Ricinus communis* L.
(Polygonaceae) *Rumex crispus* L.
(Polygonaceae) *Rumex obtusifolius* L.
(Labiatae) *Salvia lavanduloides* HBK
(Apiaceae) *Sanicula liberta* C. et S.
(Asteraceae) *Smallanthus maculatus* (Cav.) H. Robinson
(Solanaceae) *Solanum americanum* Miller
(Asteraceae) *Stevia ovata* Willd.
(Asteraceae) *Taraxacum officinale* Webb
(Asteraceae) *Tithonia diversifolia* (Hemsl.) A. Gray
(Verbenaceae) *Verbena carolina* L.
(Verbenaceae) *Verbena litoralis* Kunth

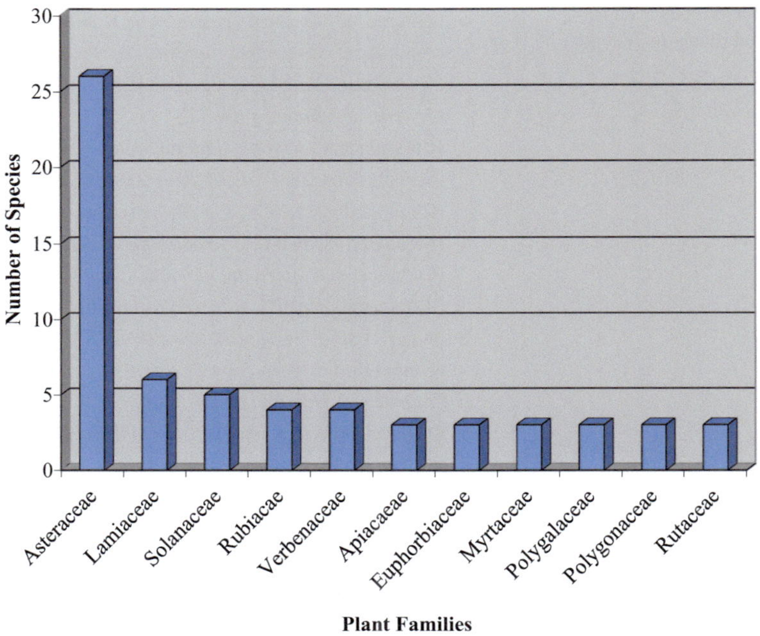

Fig. 3.15 Most Important Families of Medicinal Plants in Tenejapa

3.3 Reported Distribution of Medicinal Plants

While so far, the discussion has focused on where utilized plants are actually found, it is useful to examine local knowledge with regards to the availability and distribution of the most well known medicinal plants. These data could be used to improve management of medicinal plant resources in the context of increasing human population pressures in the Highlands, a topic I will return to in the concluding chapter. It is important to note that these data are not based on actual medicinal plant use as described previously. Instead, they are based on detailed and comprehensive interviews about medicinal plant distribution, i.e. where they are known to grow.

Long-term fieldwork in the area allowed us to use a list of 203 medicinal plants with the highest consensus as to their medicinal status (cf. Berlin and Berlin 1996). We created a travelling herbarium of dried vouchers without labels or identifying characteristics through field collections. Plants that we could not collect were borrowed from another travelling herbarium set housed at ECOSUR in San Cristóbal. Fourteen adults who were knowledgeable about medicinal plants but not specialized healers participated. They were equally divided between male and females and were from all of the major ecological zones within the municipality. Interviews were conducted in Tzeltal with the aid of collaborators.

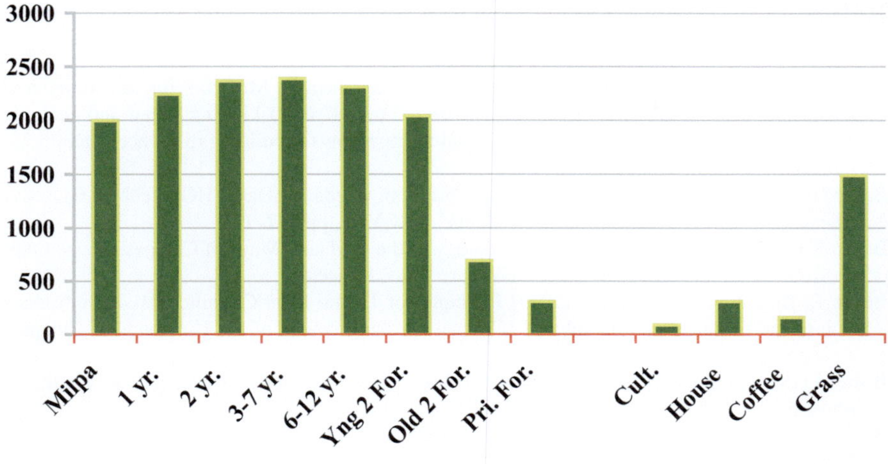

Fig. 3.16 Aggregate reported distribution of the 203 most reported medicinal plants

The participants were asked if they were familiar with and could name the plant shown to them. Then they were asked whether the plant grew in a particular successional stage or ecological zone. We did not use an exhaustive list of ethnoecological categories but rather focused on successional stages, agricultural habitats and climatic zones. Participants were also asked to note any plants that were commonly cultivated. The results are quite striking with regards to the level of generalized medicinal plant knowledge. All of the participants were highly knowledgeable about the identification of the medicinal plants, with a 98% identification rate. In total there were 2842 responses recorded. There was also a high degree of consensus regarding the availability of plants within a specific ethnoecological category and their responses correspond to where we have made actual field collections over the years. Figure 3.16 shows the reported distribution of the medicinal plants.

The highest reported concentrations appear in *unin k'inal* (3–7 year secondary growth). There is a high reported distribution of medicinal plants in disturbed areas as opposed to forested areas, especially primary forest. Overall, the results corroborate a much earlier study done with only the top 40 medicinal plants utilized in Tenejapa (Stepp 1998).

While the methodology utilized in this study can be time intensive to undertake for all involved, when compared to the time required for actual vegetation studies (i.e. obtaining the actual medicinal plant *in situ*) it is relatively quick and efficient. This methodology is highly useful for identifying plant species which may be threatened or endangered before undertaking more complex investigations. While this methodology is not a replacement for actual field studies of plant ecology, it can help guide and refine further research.

References

Alcorn JB, Toledo V (1998) Resilience resource management in Mexico's forest ecosystems: the contribution of property rights. In: Birkes F, Folke C (eds) Linking social and ecological systems: management practices and social mechanisms for building resilience. Cambridge University Press, Cambridge

Baker HG (1965) Characteristics and modes of origin of weeds. In: Baker HG, Stebbins GL (eds) The genetics of colonizing species. Academic, New York, pp 147–172

Berlin EA, Berlin B (1996) Medical ethnobiology of the Highland Maya of Chiapas, Mexico: the gastrointestinal diseases. Princeton University Press, Princeton

Berlin B, Breedlove DE, Raven P (1974) Principles of Tzeltal Plant Classification. An introduction to the botanical ethnography of a Mayan-speaking people of the Highlands of Chiapas. Academic, New York

Boster J (1987) Agreement between biological classification systems is not dependent on cultural transmission. Am Anthropol 89(4):914–920

Breedlove DE (1981) Introduction to the flora of Chiapas. California Academy of Sciences, San Francisco

Cervantes Trejo E (1997) La clasificación Tzoltzil de suelos. In: Parrra Vázquez MR, Díaz Hernández BM (eds) Los Altos de Chiapas: agricultural y crisis rural. ECOSUR, Mexico

Fabrega H, Silver D (1973) Illness and Shamanistic curing in Zinacantan. Stanford University Press, Stanford

Fairbairn JW, Suwal PN (1961) The alkaloids of hemlock (Conium maculatum L.)—II: evidence for a rapid turnover of the major alkaloids. Phytochemistry 1:38–46

Fleck DW, Harder JD (2000) Matses Indian rainforest habitat classification and mammalian diversity in Amazonian Peru. J Ethnobiol 20(1):1–36

Foster GM (1976) Disease etiologies in Nonwestern Medical Systems. Am Anthropol 78(4):773–782

Harborne JB (1993) Introduction to ecological biochemistry, 4th edn. Harcourt Brace and Company, London

Heywood V (1993) Flowering plants of the world. Oxford University Press, New York

Holland WR (1963) Medicina Maya en los Altos de Chiapas. Instituto Nacional Indigenista, Mexico City

Holling CS (1992) Cross-scale morphology, geometry, and dynamics of ecosystems. Ecol Monogr 62(4):447–502

Holm L (1978) Some characteristics of weed problems in two worlds. Proc Western Soc Weed Sci 31:3–12

Holm L, Pancho VH, Herberger JP, Plucknett DL (1979) A geographical atlas of world weeds. John Wiley and Sons, New York, p vii

Hunn E (1990) Nch'I-Wana, The Big River: Mid-Columbia Indians and their land. University of Washington Press, Seattle

Johnson J, Griffith D (1998) Visual data collection, analysis and representation. In: de Munck VC, Sobo EJ (eds) In using methods in the field: a practical introduction and casebook. Altamira Press, Walnut Creek

Martin G (1995) Ethnobotany: a methods manual. Chapman Hall, New York

Meilleur B. (1986) Alluetian ethnoecology and traditional economy: the procurement and production of plant resources in the Northern French Alps. PhD dissertation, University of Washington, Seattle

Moerman DE (1991) The medicinal flora of native North America: an analysis. J Ethnopharmacol 31:1–42

Moerman DE (1996) An analysis of the food plants and drug plants of native North America. J Ethnopharmacol 52(1):1–22

Moerman DE, Pemberton RW, Kiefer D, Berlin B (1999) A comparative analysis of five medicinal floras. J Ethnobiol 19(1):49–67

Sharer RJ (2006) The ancient Maya, 6th edn. Stanford University Press, Stanford
Shepard GH (2001) Rain forest habitat classification among the Matsigenka of the Peruvian Amazon. J Ethnobiol 21(1):1–38
Stepp JR (1998) Ethnobiology in Highland Chiapas: a look at medicinal plant distribution and conservation. Appl Geogr Dev 37:52–62
Stepp JR (2009) Informant use recall technique in ethnobotany. In: de Albuquerque UP, Hanzaki N (eds) Recent development and case studies in ethnobotany. NUPEEA, Brazil, pp 100–108
Stepp JR, Moerman DE (2001) The importance of weeds in ethnopharmacology. J Ethnopharmacol 75:19–23
Villaseñor Ríos JL, Espinosa García FE (1998) Catálogo de Malezas de México. Universidad Nacional Autónoma de México, Mexico City
Weber WA (1982) Mnemonic three-letter acronyms for the families of vascular plants: a device for more effective herbarium curation. Taxon 31(1):74–88
Whyte WF (1984) Learning from the field: a guide from experience. Sage, Newbury Park
Zimdahl RL (2018) Fundamentals of weed science, 5th edn. Academic, San Diego

Chapter 4
The Most Important Medicinal Plants of the Highland Maya

This chapter presents a botanical ethnography (cf. the subtitle of Berlin et al. 1974) of the most important medicinal plants in the municipality of Tenejapa as determined by use. Inherent in the phrase "botanical ethnography" is the interconnectedness between the local flora and the local culture. Linguistic, economic and ecological aspects of medical ethnobotany are described here for plants with the greatest use as medicines by the Tenejapa Tzeltal based on use reports over time and ethnographic interviews (Table 4.1). In addition, a reported ethnoecological distribution of these medicinal plants based on degree of human modification of the landscape and gross climatic zones is presented. Plants are listed in order of their importance. What is remarkable is that for a culture with such a large number of medicinal plants at their disposal (perhaps as high as 1500), a rather small number of species are used so frequently (five plants account for approximately 25% of all medicinal plant use). If one assumes that there is a rational basis for Maya medicine, then this is prima facie evidence that the plants are efficacious. Moreover, pharmacological studies on the bioactivity of compounds in these plants support this conclusion. Also, these plants are highly abundant in the landscape and well-known. Last, these plants treat the most commonly occurring ethnomedical conditions found in the Highlands of Chiapas.

4.1 Format of Entries

For each plant the entry is divided into two sections: Ecology, Botany, Folk Classification and Nomenclature; Medical Ethnobotany and Ethnopharmacology. In the first section the family name is given followed by a botanical description of the plant. Botanical descriptions are based on field notes and observations and a variety of sources including Berlin and Berlin (1996), Berlin et al. (1974), Morton (1981), Mabberley (2008), Ross (1999), Chevallier (1996), Strother (1999), Breedlove and Laughlin (1993), Mason and Mason (1987), and Thomson (1978).

J. R. Stepp, *Ethnoecology and Medicinal Plants of the Highland Maya*,
Ethnobiology, https://doi.org/10.1007/978-3-319-69315-6_4

Table 4.1 Most utilized medicinal plants

Name of plant	Rank	Percentage of total use
Verbena litoralis Kunth	1	11
Salvia lavanduloides Kunth	2	5
Baccharis confertoides Nesom	3	4
Tithonia diversifolia (Hemsl.) A. gray	4	4
Foeniculum vulgare L.	5	3

Authorities and proper nomenclature was verified through Mabberley (2008) and the Missouri Botanical Garden Tropicos website (www.mobot.org). This is followed by a description of the ethnoecological habitats where it is reported to grow along with its range in altitude and its distribution across gross climatic zones. If the plant grows in other places besides Chiapas this is noted. If the plant is considered a weed this is noted. At the end of each entry is the Tenejapa Tzeltal name for the plant along with any other relevant information regarding its folk taxonomy.

The medical ethnobotany and ethnopharmacology section notes what illness(es) the plant is used for and how it is prepared. Other secondary uses are also noted. It is noted if it is used by other socio-linguistic groups in different areas. If there is bioassay literature on the plant it is summarized. Any known compounds present in the plant are summarized. Ethnopharmacological descriptions are derived from a search of the NAPRALERT database (www.napralert.org; Loub et al. 1985) at the University of Illinois, Chicago although other literature is cited as well.

Following these descriptions are a photograph(s) of the plant in its natural habitat, and, if available, a botanical illustration of the plant drawn by Don Nicolás Hernández Ruíz, a Tzotzil Maya illustrator and used here with permission. Also a photograph of the herbarium voucher used in the distribution study is included.

4.2 *Verbena litoralis* Kunth

4.2.1 *Ecology, Botany, Folk Classification and Nomenclature*

This plant in the Verbenaceae (Verbena) family is the most utilized medicinal plant in Tenejapa, along with the closely related species *Verbena carolina*. The use of this plant accounted for 11% of all medicinal plant use during the study. It is an erect, branched aromatic perennial herb up to 1.5 m tall (Figs. 4.1, 4.2, and 4.3). Stems are quadrangular, sometimes minutely hairy on the angles. Leaves are nearly stemless, opposite, oblanceolate, 3–10 cm long, 1.4–4 cm wide, and short-petiolate, with serrate margins and rounded apex and are prominently veined beneath. Flowers are minute, blue to purple, tubular,5-lobed, 2.5–3 mm wide in lateral bracteal spikes 10–14 cm long. The seed capsule is quadrangular and separates into four segments, each with one seed (Berlin and Berlin 1996; Morton 1981). It commonly grows in

Fig. 4.1 *Verbena litoralis* growing in karst landscape of Nabil, Tenejapa

the *pat na* (immediate area around the house) and along trailsides and fallow fields (Fig. 4.4). It is unlikely that any household in Tenejapa is more than 1 km from a specimen of *Verbena litoralis*. This species is common throughout disturbed areas in Tenejapa and the Highlands, although less so at lower elevations (Fig. 4.5). It is common and native throughout the neotropics, extending into Arkansas, Louisiana and Texas to the north and Chile to the south. It has become naturalized in Hawaii and New Zealand. It is noted as a weed in Chiapas (Villaseñor Ríos and Espinosa García 1998). It is also considered a weed in Argentina, Colombia, Brazil, Chile, Peru, Uruguay, Venezuela, USA, Hawaii and New Zealand (Holm et al. 1979).

The plant is known by the Tenejapa Tzeltal as *yakan kulub wamal* lit. 'grasshopper's leg herb'. This is because the flowers resemble a grasshopper's leg. This name is utilized throughout the Tzeltal Highland region (Berlin and Berlin 1996). *Verbena carolina* is sometimes called this as well although it is further recognized by some Tenejapans as *sakil yakan kulub wamal* lit. 'white grasshopper's leg herb' due to the white flowers that distinguish it from the bluish-purple ones of *V. litoralis*.

4.2.2 Medical Ethnobotany and Ethnopharmacology

The leaves of this plant are brewed into a tea and drunk to treat a variety of gastrontestinal illnesses, most commonly general diarrhea (*tza'nel*). However, I noted two cases where it was used for cough (*obal*) and two cases where it was used for stomach pain (*k'ux ch'ujtil*). It is also used to treat fever and weakness and wasting type illnesses (Berlin and Berlin 1996). Its use as a phytomedicine is widespread throughout the Americas. It is used in Argentina to treat diarrhea, respiratory tract infections

Fig. 4.2 Botanical illustration of *Verbena litoralis*

and urinary tract infections (Perez and Anesini 1994), in Guatemala to treat stomach pains (Giron et al. 1991); and for menstrual disorders and as an emmenagogue in Colombia (Gonzalez and Silva 1987). Laboratory analysis shows antibacterial activity against *bacillus subtilis*, *e. coli*, *pseudomonas aeruginosa*, and *staphylococcus aureus*; antimycobacterial activity; and activity against *candida albicans* (McChesney and Adams 1985). It is also demonstrates antioxidant and antiociceptive properties (Braga et al. 2012). Compounds present include brasoside and verbenalin, monoterpenes and verbascoside, a phenylpropanoid (Umana and Castro 1990; Castro et al. 1990).

Fig. 4.3 Voucher specimen of *Verbena litoralis* used in distribution study

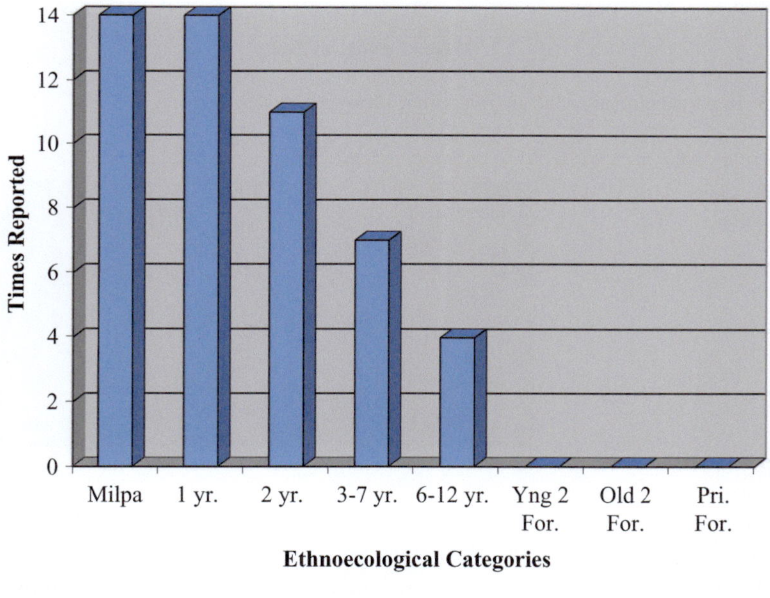

Fig. 4.4 Reported habitat distribution of *Verbena litoralis*

4.3 *Salvia lavanduloides* Kunth

4.3.1 Ecology, Botany, Folk Classification and Nomenclature

This plant in the Lamiaceae (Mint) family is the second most utilized medicinal plant in Tenejapa. The use of this plant accounts for 5% of all medicinal plant use. *Salvia lavanduloides* grows as a branched aromatic perennial spindly herb or shrub up to 1 m tall. It has a square stem and the flowers are blue, bilabiate, 0.5–1 cm long,

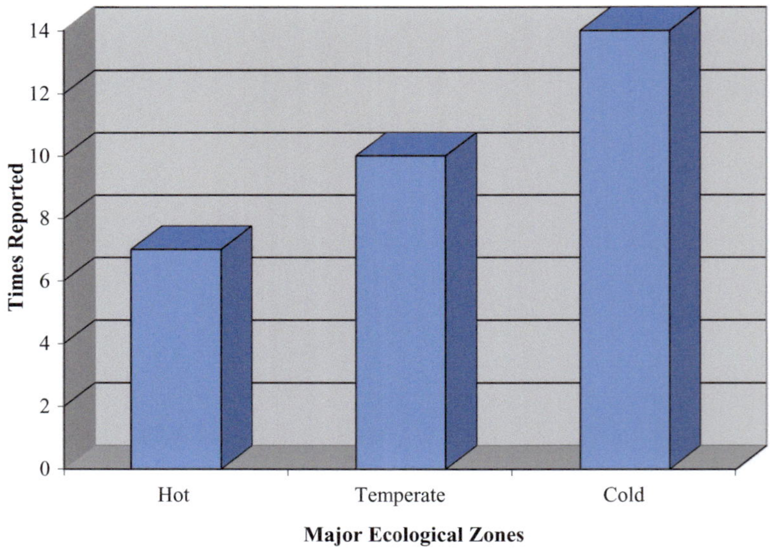

Fig. 4.5 Reported climatic zone distribution of *Verbena litoralis*

Fig. 4.6 *Salvia lavanduloides*

calyx is dull violet blue with a blue corolla (Figs. 4.6, 4.7, and 4.8). The leaves are opposite, long stemmed, 1–10 cm long, pubescent, lighter beneath, and short-petioled. The plant is readily found in highly disturbed areas, fallow fields, and young secondary forest as an understory plant with limited distribution in older secondary forest (Fig. 4.9). Two informants reported that it can be found in primary forest although it is uncommon. It does not grow in the primary forest of Matsab but three other species of *Salvia* were found there. The plant is usually found above 1800 m although its range can extend to approximately 1400 m (Fig. 4.10). *Salvia* is an abundant genera with 63 species found in Chiapas (Breedlove 1986). *S. lavanduloides* is noted as a weed in the state of Chiapas (Villaseñor Ríos and Espinosa García 1998). The plant is known to the Tenejapa Tzeltal as *ch'abakal wamal* or *ch'aalbakal wamal* (lit. 'bitter bone' herb). The significance of the plant as a medicinal is apparent in the Zinacantán Tzotzil name for this plant-*poxil obal* meaning literally cough medicine (Breedlove and Laughlin 1993).

Fig. 4.7 Close-up view of *Salvia lavanduloides* in flower

Fig. 4.8 Voucher specimen of *Salvia lavanduloides* used in distribution study

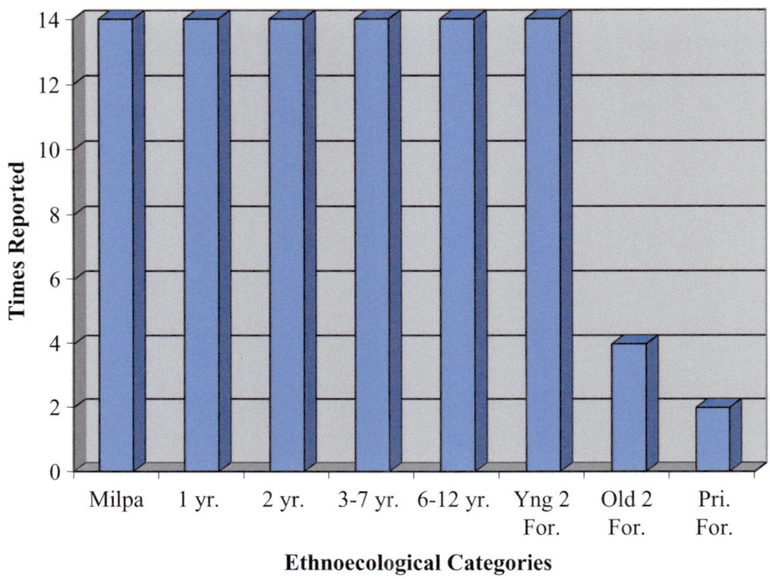

Fig. 4.9 Reported habitat distribution of *Salvia lavanduloides*

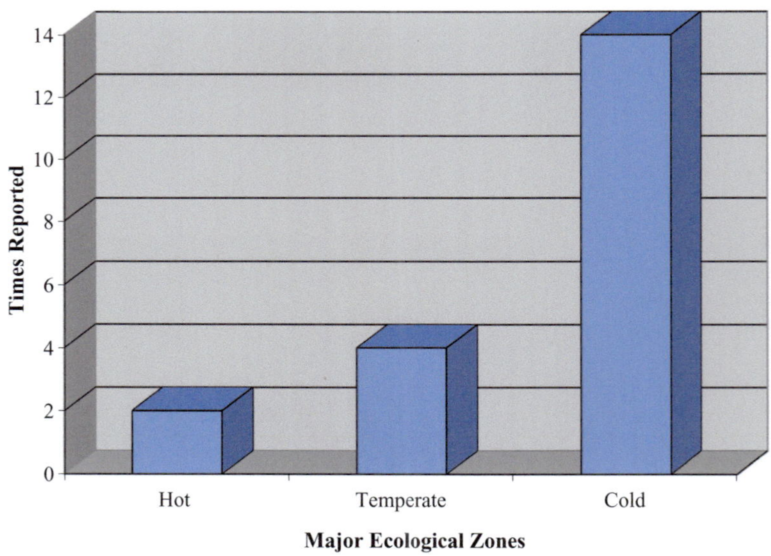

Fig. 4.10 Reported climatic zone distribution of *Salvia lavanduloides*

4.3.2 Medical Ethnobotany and Ethnopharmacology

The plant is used to treat generalized cough (*obal*) with one reported use for diarrhea (*tza'nel*). It is the most utilized cure for respiratory illnesses. The leaves and stems are boiled in water to make a tea. Bioactive triterpenoids are found in *Salvia lavanduloides* (Topçu 2006). The worldwide use of plants from this genus for medicinal purposes suggests that this species merits further detailed study of its potential biological activity. The plant is known to contain the following flavonoids: santin, eupatorin, flavone,5-hydroxy-3′-4′-6–7-tetrame, flavone, 5-hyfroxy-3′-4′-6-tetrametho, and glucoferide; the following diterpene: horminone; the following triterpenes: oleanolic acid, ursolic acid; and the following diterpenes: salvianduline A, salvianduline A, salvianduline B, salvianduline C, and salvianduline D (Rodriguez et al. 1974; Ortega et al. 1991; Maldonado et al. 1992).

4.4 *Baccharis confertoides* G.L. Nesom

4.4.1 Ecology, Botany, Folk Classification and Nomenclature

Baccharis confertoides Nesom is in the Asteraceae (Sunflower) family. It is the third most utilized medicinal plant in Tenejapa. The type of *Baccharis vaccinioides* Kunth is synonymous with that of *Baccharis conferta* Kunth and the name *B. vaccinioides* has long been misapplied to plants that are now called *B. confertoides* (Nesom 1990; cf. Berlin and Berlin 1996). The plant grows as a shrub or treelet generally 2–6 m tall (Figs. 4.11, 4.12, 4.13, and 4.14). Leaves are alternate, obovate or oblancelolate to elliptic, 1–3.5 cm long, 0.3–1.5 cm wide. The margins are entire to minutely spiny-toothed. Apices are acute, faces are glabrous, usually glutinous or vernicose. Infloresences are white and in small heads in 3–8 compact terminal clusters from 8 to 12 cm long. Florets number 20–40. It is a common plant on the central plateau of Chiapas although it is rarely found in the Sierra Madre Mountains and the Northern Highlands. The plant appears to thrive in karst topography, perhaps due to the high limestone content of the soil. It is commonly found along trailsides and roads and in openings in secondary forests and fallow fields (Fig. 4.15). It often forms pure stands in disturbed areas. The plant generally grows above 2000 m (Fig. 4.16). Other *Baccharis* species are noted as weeds through North America and

Fig. 4.11 *Baccharis confertoides*

Fig. 4.12 *Baccharis confertoides*

Fig. 4.13 Voucher specimen of *Baccharis confertoides* used in distribution study

the neotropics but *B. confertoides* is not listed as a weed (Holm et al. 1979). The plant is known in Tenejapa and throughout the highlands as ***mes te'*** (lit. Broom tree) because the branches of the plant are gathered together into a makeshift broom for sweeping.

4.4.2 Medical Ethnobotany and Ethnopharmacology

In Tenejapa this plant is used to treat diarrhea (***tza'nel***) and cough (***obal***). When used to treat diarrhea it is much more likely to be used as an admixture rather than a primary plant. However when used for cough it is not taken with any other plants. For both illnesses, the leaves of the plant are stripped off of the stems and then

Fig. 4.14 Botanical illustration of *Baccharis confertoides*

boiled in water. While the Tzotzil Maya also use the plant in this manner it is unknown if it is used medicinally outside of the highlands of Chiapas. In a methanolic extract it shows strong spasmolytic effect and also cytotoxic activity on KB and P388 cell cultures (Berlin and Berlin 1996). A slight antimicrobial activity was noted against *Staphylococcus aureus* (Berlin and Berlin 1996). Numerous flavonoids have been isolated from B. confertodies (Wollenweber et al. 1986). Phytochemical analysis of closely related species reveals a range of bioactivity (Freire et al. 2007).

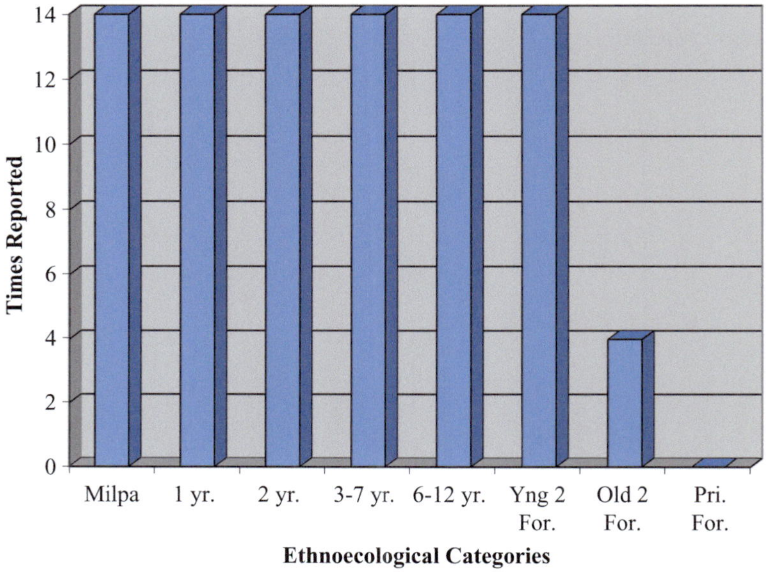

Fig. 4.15 Reported habitat distribution of *Baccharis confertoides*

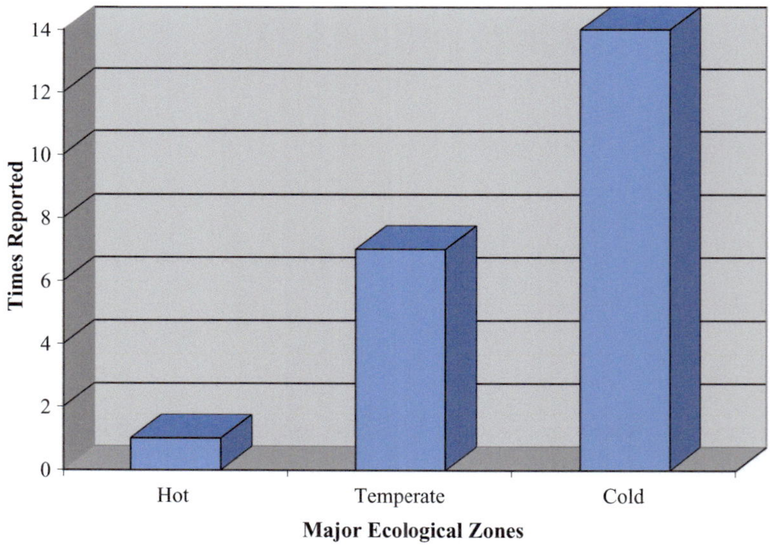

Fig. 4.16 Reported climatic zone distribution of *Baccharis confertoides*

4.5 *Tithonia diversifolia* (Hemsl.) a. Gray

4.5.1 Ecology, Botany, Folk Classification and Nomenclature

Tithonia diversifolia (Hemsl.) A. Gray is in the Compositae (Sunflower) family. It is the fourth most utilized medicinal plant in Tenejapa. The plant is an herbaceous or shrubby perennial up to three meters tall. The petioles are 2–6 cm long, winged. The leaf blades are lance-ovate to rhombic, deltate or pentagonal, 10–20 cm long, 6–18 cm wide and densely pubescent on both surfaces. The flowers are bright yellow to orange with ray florets 12–13 in number and appear mostly in September to October (Figs. 4.17, 4.18, and 4.19). The plant is found throughout disturbed areas in Tenejapa and at all elevations, although it is reported to be slightly more prevalent at lower altitudes (Figs. 4.20 and 4.21). The native distribution is throughout Southern Mexico, into Central America and down into Colombia, Venezuela and Ecuador. It is also naturalized in the Old World tropics. It is noted as a weed in Chiapas and also in Australia, Indonesia, Fiji, Thailand, the continental United States, Hawaii and Mauritius (Holm et al. 1979). The plant is known in Tenejapa as *ch'akil* (unanalyzable constituent).

4.5.2 Medical Ethnobotany and Ethnopharmacology

In Tenejapa this plant is used to treat diarrhea (***tza'nel***) and cough (***obal***) and abdominal pain (*k'uxch'ujtil*). Berlin and Berlin (1996) note that it is one of the principal species used by the Tzotzil in the treatment of abdominal pain. It is not reported to be used for cough in Berlin and Berlin (1996) but in Tenejapa this accounts for 50% of its use. The leaves are boiled in water and a tea is made to treat the illness. It is used to treat malaria in Guatemala (Calzada and Ciccio 1978). In Kenya it is used to treat gastrointestinal problems (Johns et al. 1995). In Thailand it is used to treat stomach pain (Anderson 1986). Studies have demonstrated the

Fig. 4.17 *Tithonia diversilfolia* in flower

Fig. 4.18 Botanical illustration of *Tithonia diversilfolia*

Fig. 4.19 Voucher specimen of *Tithonia diversifolia* used in distribution study

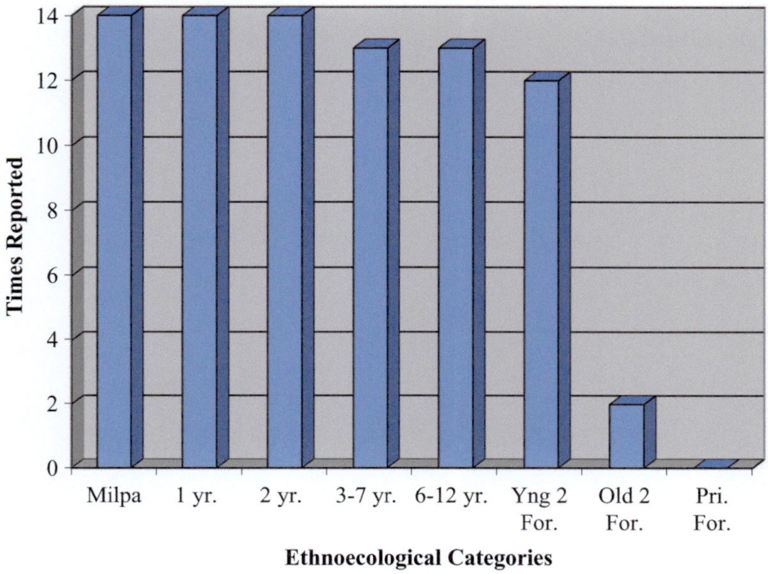

Fig. 4.20 Reported habitat distribution of *Tithonia diversifolia*

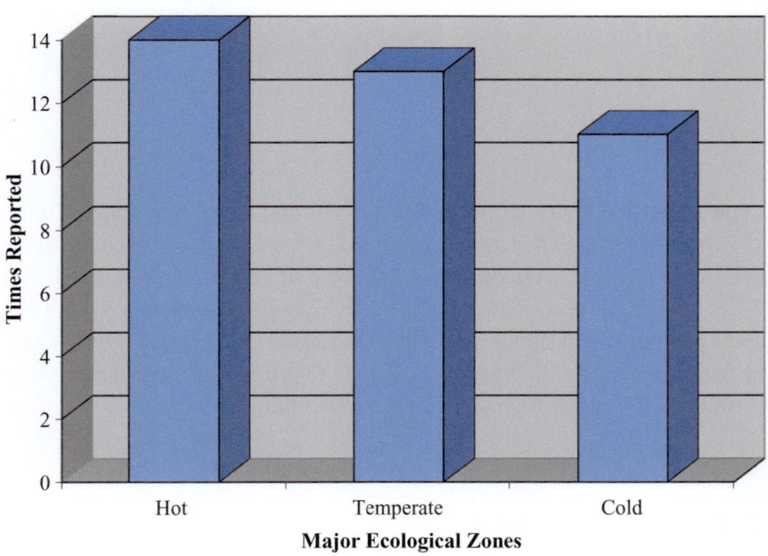

Fig. 4.21 Reported climatic zone distribution of *Tithonia diversifolia*

Fig. 4.22 Botanical illustration of *Foeniculum vulgare*

plant's spasmolytic activity (Kambu et al. 1990) and its cytotoxicity (Mungarulire 1991). Numerous terpenoids (monterpenes and sesquiterpenes) such as tagitinin A, B, C, and ocimene have been isolated from the plant (Lamaty et al. 1991; Schuster et al. 1992). In 2012, two new compounds 6″-*O*-*β*-D-apiofuranosyl-trichocarpin (15) and 1-heptade-4,6-diyne-3,10,16,17-tetraol-3-*O*-*β*-D-glucopyranoside were isolated from this species (Zhao et al. 2012). There is also significant phytotoxic activity found in *Tithonia diversilfolia* (Miranda et al. 2015).

4.6 *Foeniculum vulgare* L.

4.6.1 Ecology, Botany, Folk Classification and Nomenclature

Foeniculum *vulgare* L., commonly known as fennel, is in the Umbelliferae (Carrot) family. It is the fifth most utilized medicinal plant in Tenejapa. The plant is a coarse aromatic herb with a biennial or perennial root with an annual erect green stem up

Fig. 4.23 Voucher specimen of *Foeniculum vulgare* used in distribution study

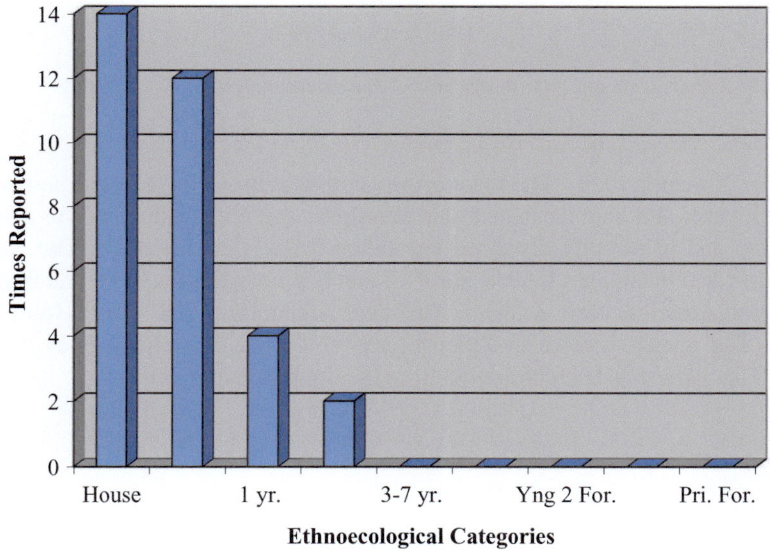

Fig. 4.24 Reported habitat distribution of *Foeniculum vulgare*

to 2 m with finely divided, glabrous leaves up to 3 cm long, alternate, feathery and composed of deep green, forked, threadlike segments (Figs. 4.22 and 4.23). The flowers are small, yellow and in terminal flat-topped umbels up to 15 cm across. The seed is aromatic, oblong, ridged 5 mm long, 1.5–2 mm wide. The plant is native to the Mediterranean but is now cultivated worldwide. It is considered to be one of the most important and widely used medicinal plants worldwide and treats at least forty different disorders (Badgujar et al. 2014). It was introduced in Chiapas shortly after contact, with historical literature noting that it was introduced into the nearby Tzotzil municipality of Zinacantán in 1545 by Fray Jordan de Piamonte from

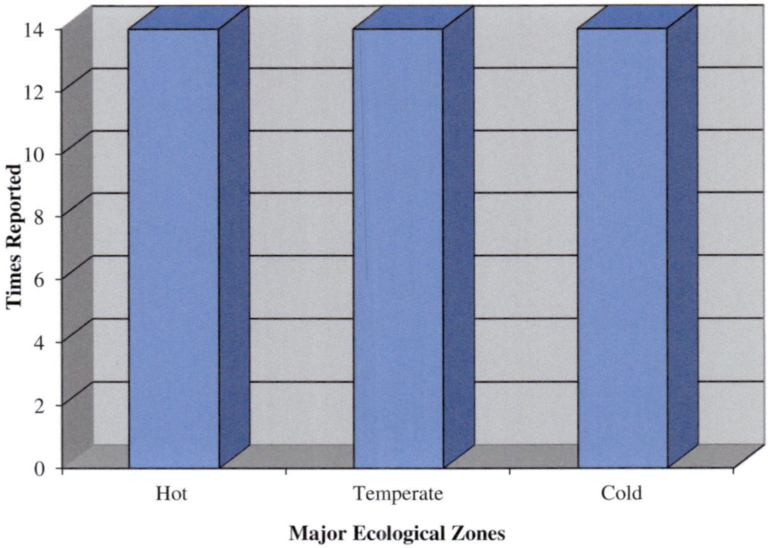

Fig. 4.25 Reported climatic zone distribution of *Foeniculum vulgare*

Oaxaca (Ximénez 1929). The plant grows at all elevations in Tenejapa and is usually cultivated although it can be found in milpas, where it is maintained as a volunteer crop and in recently disturbed vegetation (Figs. 4.24 and 4.25). The plant is noted as weed in Chiapas (Villaseñor Ríos and Espinosa García 1998) and throughout South America, New Zealand, Australia, Afghanistan and Israel (Holm et al. 1979). The plant is known in Tenejapa and in Tzotzil municipalities as *inojo* or *inajo*. This name has been borrowed from the Spanish name *hinojo*.

4.6.2 Medical Ethnobotany and Ethnopharmacology

The plant is used primarily to treat epigastric pain, more commonly known as colic and in Tzeltal, *sk'ajk'al o'tanil*. It is less commonly used for heartburn (*k'ux o'tonil*). The leaves are gathered and then boiled into a tea and occasionally the root is used as well. Fennel is a significant medicinal plant worldwide and it use traces back at least to ancient Greek civilization. It is included in the Physician's Desk Reference for Herbal Medicines and was also included in the German government's commission E list of safe and effective medicinal plants (Thomson Healthcare 2007).

A wide range of studies have shown the effectiveness of fennel in the treatment of epigastric pain (e.g. Giordano and Levine 1989). The essential oil of the seeds contain D-pienen, camphene, D-phellandrene, dipenene, fenchone, methyl chavicol, aldehydes and anisic acid (Furia and Bellanca 1971). It has been shown to be antifungal and antioxidative (Singh et al. 2006).

References

Anderson EF (1986) Ethnobotany of hill tribes of northern Thailand II. Lahu medicinal plants. Econ Bot 40(4):442–450

Badgujar SB, Patel VV, Bandivdekar AH (2014) *Foeniculum vulgare* mill: a review of its botany, phytochemistry, pharmacology, contemporary application, and toxicology. Biomed Res Int 2014:842674

Berlin EA, Berlin B (1996) Medical ethnobiology of the Highland Maya of Chiapas, Mexico: the gastrointestinal diseases. Princeton University Press, Princeton

Berlin B, Breedlove DE, Raven P (1974) Principles of Tzeltal plant classification. An introduction to the botanical ethnography of a Mayan-speaking people of the highlands of Chiapas. Academic, New York

Braga VF, Mendes GC, Oliveira RTR, Soares CQG, Resende CF, Pinto LC, de Santana R, Viccini LF, Raposo NRB, Peixoto PHP (2012) Micropropagation, antinociceptive and antioxidant activities of extracts of Verbena litoralis Kunth (Verbenaceae). An Acad Bras Cienc 84(1):139–148

Breedlove DE (1986) Listados Florísticos de México. IV. Flora de Chiapas. Instituto de Biología, UNAM, México

Breedlove DE, Laughlin R (1993) The flowering of man: a Tzotzil botany of Zinacantán. Smithsonian Institution Press, Washington, DC

Calzada JG, Ciccio JF (1978) Isolation of Tirotundin from Tithonia diversifolia. Revista de Latin Americana Quimica 9:202

Castro O, Umana E, Herrera ML (1990) Biological and chemical potential of Verbena litoralis: a medical plant used in Costa Rica as an antidiarrhea agent. Quim Nova 13(4):310–311

Chevallier A (1996) The encyclopedia of medicinal plants. Dorling Kindersley Limited, London

Freire S, Urtubey E, Giuliano AD (2007) Epidermal characters of Baccharis (Asteraceae) species used in traditional medicine. Caldasia 29:23–38

Furia TE, Bellanca N (1971) Fenaroli's handbook of flavor ingredients. The Chemical Rubber Co, Cleveland

Giordano J, Levine PJ (1989) Botanical preparations used in Italian folk medicine: possible pharmacological and chemical basis of effect. Soc Pharmacol 3(2):83–110

Giron LM et al (1991) Ethnobotanical survey of the medicinal flora used by the Caribs of Guatemala. J Ethnopharmacol 34(3):173–187

Gonzalez F, Silva M (1987) A survey of plants with antifertility properties described in the South American Folk Medicine. Abstracts of Princess Congress I, Bangkok

Holm L, Pancho VH, Herberger JP, Plucknett DL (1979) A geographical atlas of world weeds. Wiley, New York

Johns T et al (1995) Anti-giardial activity of gastrointestinal remedies of the Luo of East Africa. J Ethnopharmacol 46(1):17–23

Kambu K et al (1990) Antispasmodic activity of extracts used in Zaire. Ann Pharm Fr 48(4):200–208

Lamaty G et al (1991) Aromatic plants of tropical Central Africa. J Essent Oil Res 3(5):399–402

Loub W et al (1985) NAPRALERT: computer handling of natural products research data. J Chem Inf Comput Sci 25:99–103

Mabberley DJ (2008) The plant book: a portable dictionary of the vascular plants, 3rd edn. Cambridge University Press, Cambridge

Maldonado E et al (1992) Salviaduline C, A 5,6-secoclerodane Diterpenoid from Salvia lavanduloides. Phytochemistry 31(1):217–220

Mason CT, Mason PB (1987) A handbook of Mexican roadside flora. The University of Arizona Press, Tucson

McChesney JD, Adams RP (1985) Co-evaluation of plant extracts as petrochemical substitutes and for biologically active compounds. Econ Bot 39(1):74–86

Miranda MA, Varela RM, Torres A, Molinillo JMG, Gualtieri SCJ, Macías FA (2015) Phytotoxins from Tithonia diversifolia. J Nat Prod 78(5):1083–1092

Morton J (1981) Atlas of medicinal plants of middle America: Bahamas to Yucatan. Charles C. Thomas, Springfield

Mungarulire J (1991) A phytochemical investigation of medicinal plants of the compositae from Rwanda. Herba Hung 29(3):73–78

Nesom GL (1990) Two new species of Mexican Baccharis (Asteraceae: Astereae). Phytologia 68(6):32–39

Ortega A et al (1991) Salviandulines a and B: two secoclerodane diterpenoids from salvia lavanduloides. Phytochemistry 30(10):3357–3360

Perez C, Anesini C (1994) Inhibition of Pseudomonas aerguinosa by argentinean medicinal plants. Fitoterapia 65(2):169–172

Rodriguez J et al (1974) Flavanoids of Mexican plants: isolation and structure of santin and glucoferide. Revista Latinoamerica Quimica 5:41–53

Ross IA (1999) Medicinal plants of the world: chemical constituents, traditional and modern medicinal uses. Humana Press, Totowa

Schuster A et al (1992) Sesquiterpene lactones from two Tithonia species. Phytochemistry 31(9):3139–3141

Singh G, Maurya S, de Lampasona MP, Catalan C (2006) Chemical constituents, antifungal and antioxidative potential of Foeniculum vulgare volatile oil and its acetone extract. Food Control 17(9):745–752

Strother JL (1999) The Flora of Chiapas part 5: Compositae-Heliantheae s.l. California Academy of Sciences, San Francisco

Thomson W (1978) Medicines from the earth: a guide to healing plants. Harper and Row, San Francisco

Thomson Healthcare (ed) (2007) PDR for herbal medicines, 4th edn. Thomson Reuters, New York

Topçu G (2006) Bioactive triterpenoids from Salvia species. J Nat Prod 69(3):482–487

Umana E, Castro O (1990) Chemical constituents of verbena litoralis. Int J Crude Drug Res 28(3):175–176

Villaseñor Ríos JL, Espinosa García FE (1998) Catálogo de Malezas de México. Universidad Nacional Autónoma de México, Mexico

Wollenweber E et al (1986) Flavonoids and terpenoids from the exudates of some Baccharis species. Naturforsch Ser C 41(2):87–93

Ximénez F (1929) Historia de la Provincia de San Vicente de Chiapas y Guatemala de la Orden de Predicadores (3 volumes). Tipografía Nacional, Guatemala

Zhao GJ, Xi ZX, Chen WS, Li X, Sun L, & Sun LN (2012) Chemical constituents from Tithonia diversifolia and their chemotaxonomic significance. Biochem Syst and Ecol 44:250–254

Chapter 5
But What About the Forest? Mature Forests and Highland Maya Medicinal Plants

Beginning in the late 1980s, there was a flourishing of public interest in the conservation of the rapidly diminishing mature tropical forests of the world. While numerous arguments were developed to justify their conservation, those involving economic arguments proved to be most effective with the general public (Kempton et al. 1995). Mature tropical forests were valorized for their potential in providing new food crops and medicinals (see Table 5.1). Valuation studies claimed that tropical mature forests were worth more intact as a source of phytomedicinals and novel compounds for pharmaceuticals than for their actual timber value or other land uses (Mendelshohn and Balick 1995; Principe 1991).

At first glance it appears that this argument has merit. For example, Farnsworth and Soejarto (1991) found that of the 95 plant species currently used for prescription drugs, 39 are plants said to originate in and around tropical forests. No data are provided by Farnsworth and Soejarto as to how many of the species are tropical mature forest obligates (those species that grow *only* in mature forest). Further investigation finds that of these 39 species, 13 of them are considered to be weeds (Stepp 2004).

By the early 1990s the assumption that tropical mature forests were rich in potential pharmaceuticals was widely shared by the public at large (Kempton et al. 1995). At around the same time, there was increased public recognition of ethnobotany as a discipline that could aid in the quest for new pharmaceuticals by suggesting potential candidate plant species for bioassay based on their ethnobotanical use by specialized healers or shamans (Table 5.2). The linkage of ethnobotany with the search for new drugs and conservation of mature tropical forests had a significant impact on the field.

At least within the natural sciences, the field of ethnobotany had long languished as an obscure discipline, afforded little respect or interest by other scientists and was mostly unknown to the general public. The field had fared somewhat better in the social sciences and within anthropology where insights and theoretical developments dealing with ethnobotanical classification had been influential in the development of ethnoscience and cognitive anthropology (e.g. Berlin et al. 1974, for a good

J. R. Stepp, *Ethnoecology and Medicinal Plants of the Highland Maya*, Ethnobiology, https://doi.org/10.1007/978-3-319-69315-6_5

Table 5.1 Utilitarian arguments for mature tropical forest conservation

"Most of the plants that have not been analyzed and tested grow in the tropics—a large fraction of them in the rain forests. Insofar as the forests are being destroyed, species are being lost that might yield useful medicine (Abelson 1990: 513)."
"There exists no shortage of 'wonder drugs' waiting to be found in the rain forests, yet we in the industrialized world are woefully ignorant about the chemical-and, therefore medicinal-potential of most tropical plants (Plotkin 1993: 7)."
"In order to slow the rapid destruction of tropical forests and the concurrent loss of biodiversity, it is critical to identify and quantify the benefits of conserving the remaining standing forests... the existence of undiscovered pharmaceuticals for modern medicine has often been cited as one of the most important reasons to protect tropical forests (Mendelshohn and Balick 1995: 223)."

historical account see D'Andrade 1995). The rigor demonstrated in these sub-fields subsequently influenced empirically oriented social scientists interested in developing more robust methods and field techniques.

Regardless, public interest in ethnobotany was almost non-existent until the early 1990s when ethnobotanists began to ride the coattails of the growing movement to conserve mature tropical forest. In many ways, the results have been positive. Ethnobotany programs have sprung up in departments around the world. In many universities, there has been growing demand for ethnobotany classes at the undergraduate and graduate level. Traditional peoples are increasingly doing their own ethnobotanical research.

Interestingly enough, the assumption that mature tropical forests were a significant habitat for indigenous peoples to find medicinal plants was simply taken for granted (Voeks 2004). Like the popular myth of the 'hundreds of words for snow in Eskimo (Inuit),' tropical rainforests as untapped stores of new pharmaceuticals became a 'just so' type of common knowledge. Voeks (2018) connects the origins of this belief all the way back to edenic narratives arising out of Judeo-Christian culture that were amplified in the 1500s with the European supposed 'discovery' of the New World. Clearly this narrative of jungle medicine resonated with R.E. Schultes, often dubbed the father of economic botany. His influential work in the Amazon from 1941 to 1953 led to a remarkable 24,000 plant collections with 1500 species noted as medicinals or poisons by the Amazonian indigenous groups with whom he worked (Plotkin 1993). However, it is unclear exactly how many of these plants were mature forest obligates and perhaps more importantly, the frequency with which they were collected by indigenous peoples in mature forest habitats.

Systematic studies of *actual* plant use and procurement have been rare in ethnobotany. However, there are several studies that suggest mature tropical forest is not nearly as significant a habitat for indigenous peoples to obtain medicinal plants as one could be led to believe by the statements in Table 5.2. Voeks (1996) found that healers along the Atlantic coastal tropical forest of Brazil overwhelmingly preferred secondary forest and disturbed habitats. Using carefully controlled plots he evaluated the medicinal plant potential in different types of habitats and found that the majority of species used medicinally show up in disturbed areas. He also noted that the most common life form for medicinals was herbaceous plants and that trees were the least common.

Table 5.2 The linkage of utilitarian arguments for mature tropical forest conservation with the promotion of ethnobotany

" 'We all had a mantra that we had to save the rain forest because it was a repository of natural drugs,' said Wade Davis, who studied ethnobotany at Harvard. (Christensen 1999)."
"Tropical forest peoples represent the key to understanding, utilizing and protecting tropical plant diversity…a single Amazonian tribe of Indians may use over 100 species of plants for medicinal purposes alone (Plotkin 1991: 57)."
"Local forests are the source of the plants processed into therapies used in traditional medical systems (Balick and Mendelsohn 1992 citing Balick 1990)"
"How much do the tropical rain forests contribute to the mature health care of man? With such a tremendous store of biotic richness, certainly one expects them to be a storehouse, and a source, of potentially important medicinal drugs (Farnsworth and Soejarto 1991: 29)."
"This is the last generation of scientists who have the opportunity to study indigenous groups and their interaction with the rain forest." says Michael Balick of the New York Botanical Garden. "Logic says that among the tens of thousands of plants that haven't been studied there must be something of value (Jackson 1989: 97)."
"Concentrating on those plants that Amazonian tribal peoples by experimentation over the centuries have found to be bioactive could provide chemists with a kind of short cut. The value of learning from the perspicacity of many aboriginal peoples about the biodiversity of plants–those which they use and those which they know only as elements of the forest–should not be overlooked. Ethnobotanical studies devoted to this aspect of field research should be given greater and immediate attention in view of the unchecked and wanton destruction of vast areas of the world's forests. (Schultes and von Reis 1995: 13)."
"How do ethnobotanists choose the societies they study?...First, the societies should be located in a floristically diverse area, such as a tropical rain forest. Such diversity dramatically increases the number of plants available; it thus enhances the likelihood that plants with pharmacologically active molecules will be pressed into service. (Cox and Balick 1994: 84)."
"One of the many aspects of the Amazon that are difficult to comprehend is the intimate relationship between the forest Indian and his ambient vegetation…The widespread belief that his 'secrets,' especially his uncanny knowledge of the plants and animals, must be pried from him is, in general, a falsehood. (R.E. Schultes in Plotkin 1993: viii)."
"The conservation community was beginning to wake up and pay attention to the importance of ethnobotany for rain forest preservation and utilization…Somehow both plants and indigenous peoples had to be factored into the equation…Ethnobotany takes a holistic approach to conservation and proves the vital importance of the role indigenous people can play in the ongoing struggle to protect, use and sustain the rain forest…The beauty of ethnobotany is that it brings people into the forest picture, showing that tribal peoples can help provide us with answers on the best ways to use and protect the forest. (Plotkin 1993: 272–273)."
"Michael Balick of the New York Botanical Garden just returned from the interior of Belize, where he worked with a well known *curandero*, or healer, who uses herbal remedies to treat illness. As the forests disappear, however, so do sources such as Balick's medicine man. 'These people are encyclopedias of local flora and they're not being replaced'. (Booth 1987: 970)."

Other case studies corroborate Voeks. Posey's research (1984) with the Kayapó in Brazil found extensive use of disturbed habitat for medicinal plant procurement. In what he terms "anything but abandoned" fields he found that 94% of the 368 species collected were of medicinal importance. Posey found that the Kayapó utilize almost no medicinal plants from the mature forest.

A study by Caniago and Siebert (1998) did not specifically look at habitat preference but looked at medicinal plant distribution in different ecological zones for a community in Indonesia. They found that while there was some use of the mature forest for medicinals (42 species), by and large, secondary growth habitats were richer in medicinals (227 species). Alcorn's work (1984) with the Huastec (Teenek) Maya in the state of Veracruz, Mexico also found that medicinal plants appear to be collected more often from disturbed habitats. Other indigenous groups in Mexico appear to follow similar patterns. Both the Mixe and Zapotec demonstrate an overwhelming preference for selecting medicinal plants in disturbed habitats with over 80% of plants used as medicine occurring in non-forested disturbed areas (Frei et al. 2000).

Even those who have argued that mature forest is the most important habitat for traditional people to gather medicinal plants have demonstrated the significance of disturbed habitats in their own research (Stepp and Moerman 2001). For example, a perusal of Arvigo and Balick's book *Rainforest Remedies* (1993), an account of the most utilized medicinal plants in Belize based on their work with 24 specialized healers, finds that the vast majority of the "100 healing herbs" of the rainforest are found in disturbed habitats such as roadsides and fields. Of the 100 plants listed in it, only ten are said to grow exclusively in forests.

A more comprehensive work by Balick and Arvigo (2015) contains a larger number of species. However, for most species listed, habitat information is not provided, making it difficult to determine where plants might be procured. The authors make clear though they continue to be concerned with the effects of tropical deforestation on availability of medicinal plants. They mention a "dramatic reduction in supply of certain primary and secondary forest species used as medicinal plants available to the traditional healers…it does impact a large number of "bush" species widely utilized in Belize" (Balick and Arvigo 2015: 37). It would be useful to test actual medicinal plant use in lowland Belize in order to get a clear picture of exactly how important primary and secondary forests are. If, in fact, Balick and Arvigo are correct, Belize would be a significant exception compared to many other parts of the world, including nearby Chiapas.

In general though, it appears that in the enthusiasm to conserve mature tropical forest and promote ethnobotany, an unsupported assumption is being made. A colleague once told me that he was pulled aside at an ethnobotany academic conference and taken to task for reporting his data showing the abundance of use of medicinal plants within the Asteraceae family (a family characterized primarily by herbaceous plants, many with weedy qualities). His interrogator did not find fault with his work but rather expressed his concern that such findings were contrary to the public image of indigenous peoples and medicinal plants. While paradigm shifts in science are often unpleasant and characterized by divisiveness and emotion (Kuhn 1970) it is important for ethnobotanists to let the data speak for itself. "But the rhetoric (*about the rainforest being a repository of new drugs-JRS*) ran away from reality, and the whole thing has backfired. We have not found new drugs. And the fact that the idea is deeply flawed is never questioned (Wade Davis quoted in Christensen 1999)."

Bioprospecting for new pharmaceuticals from the rainforest continues to this day (Skirycz et al. 2016; Calderon et al. 2001) and environmental organizations continue to make a linkage between tropical forest conservation and medicines. For example, the Rainforest Trust website includes an article entitled "Owed to Nature: Medicines from Tropical Forests" (Torrence 2012). *Earth Talk* (2017), a feature of *E-The Environmental Magazine*, recently published an article entitled "Tropical Rainforests: Nature's Medicine Cabinet" that was subtitled "preserving rainforests may be a matter of life or death."

5.1 Medicinal Plants and Mature Forest: The Tzeltal Maya Case

When I first began fieldwork in Highland Chiapas I expected that there would be some use of the little remaining mature forest for medicinal plants (see Chap. 3 for a full description and definition of mature forest in Highland Chiapas). However, over time, I learned that this is not the case. To summarize the findings:

1. There was no use of mature forest (*ja'mal*) by any of the participants in the study as a source of medicinal plants
2. There was no use of old secondary forest (*tojol k'inal*) by any of the participants in the study
3. Young secondary forest (*te'tikil*) was utilized only 16% of the time as a habitat to obtain medicinal plants.
4. The average amount of time spent procuring medicinal plants for a specific treatment is 16 min, meaning that an area with a radius of less than 1 km around the house is the mature zone in which to collect medicinal plants.

On first glance it appears that these data clearly demonstrate that more disturbed habitats will be visited more frequently to obtain medicinal plants. Secondary forest is abundant in Tenejapa and almost every household is within 1 or 2 km of secondary forest. Yet it is used infrequently and old secondary forest and mature forest are not used at all.

This argument can be attacked on the grounds that there is little mature forest left in Tenejapa or in the Highlands at large (Figs. 5.1 and 5.2). It was estimated in 1982 that there was less than 398 km^2 of mature forest remaining in Northern Chiapas from an "original" area of approximately 1025 km^2 (Bubb 1991). However, the greatest losses have occurred in the surrounding municipalities of Tenejapa. For example, there is nothing remaining of what was once 864 km^2 of mature forest in San Andrés Larráinzar. The adjacent mature forest of Tzontehuitz is now only 5 ha, from an original area of 102 km^2. Significantly, the mature forests of Tenejapa were not included in the study by Bubb (1991) probably due to their small size and relative inaccessibility (neither lie near paved roads). It could be that there are other overlooked small patches of mature forest in other municipalities as well.

Fig. 5.1 Under the mature forest canopy in Matzab, Tenejapa

Fig. 5.2 Bromeliads growing in mature forest of Matzab in Tenejapa

I estimate that a little more than 5% of the land in Tenejapa is mature forest. I obtained this figure by utilizing a grid of 1 km^2 increments from the 1:50,000 topographical map entitled Oxchuc.E15D52 (INEGI 1990). In travelling throughout the municipality I have found mature forest in three areas: The two hills adjacent to the community of Matzab (Fig. 5.3), The hilly area adjacent to the community of Ach'lum, and a narrow ridge between the communities of Majosik' and Oxeb Wits. Because of the high edge to area ratio of this latter ridge I do not include it in this estimate. However, the mature forests of Matzab and Ach'lum are relatively intact although there has been some selective logging and one of the slopes of Matzab was accidentally burned in the early 2000s but has subsequently regenerated (albeit as secondary forest). Both areas lie between 2400 and 2600 m. The combined area amounts to approximately 5 km^2 (Matzab is approximately 2 km^2, Ach'lum is approximately 3 km^2). Given that the total area of the municipality is 99 km^2, a little more than 5% then is mature forest.

Another basis to challenge the findings in Chap. 3 comes from the fact that none of the households that participated in the research were near these areas of mature forest. The closest community to the forests of Matzab and Ach'lum was Balum Kanal, yet it was still 6–8 km away. Majosik' lies only 2–3 km from the narrow strip of mature forest to the northeast of the community. Yet even this shorter distance is still greater than the 1 km radius where the majority of medicinal plant collection takes place. Taking this into account, we initiated a study to examine use of the mature forest for medicinal plants by people living in communities close to mature forest.

Fig. 5.3 The twin mature forest hills of Matzab

5.2 Methodology

Based on reconnaissance trips around the municipality I was familiar with the communities that were adjacent to the mature forests of Matsab and Ach'lum. These communities are Matsab, Ach'lum, Santa Rosa, and Banabil. I sought out individuals from these communities who were knowledgeable about medicinal plants and willing to participate in the study. Ten individuals participated in the study (eight males and one female) who ranged in age from 35 to 50 years old (the average age was 41.2 years old). I did not interview individuals older than 50 because older individuals are less likely to gather medicinal plants and instead rely on their children and grandchildren to obtain them.

Previous research in the mature forest of Matsab identified eight plant species that were considered medicinal and were mature forest obligates (Casagrande 2004). Using voucher specimens of these eight plants (Figs. 5.4, 5.5, 5.6, 5.7, 5.8, 5.9, 5.10, and 5.11) I asked the informants if they knew the plant, and its uses (Table 5.3). They were then asked if they had ever used the plant themselves by employing a use recall methodology similar to that described in Chap. 3. The same questions were asked, except that informants were requested to note any illnesses they could remember and what the treatments were, rather than just the illnesses in the previous week.

5.3 Findings

Informants were very familiar with the eight mature forest obligate species, and were able to correctly identify them 97% of the time. Responses varied regarding the uses of the plants (Table 5.3). While plants such as *Aechmea nudicaulis*, and *Sedum praealtum* were unanimously reported as medicinals, consensus was much

Table 5.3 Mature forest obligate species with potential use as medicinals

Scientific name	Family	Tzeltal name	Reported as medicinal (%)
Aechmea nudicaulis L.	BML	**yaxtinin**	100
Sedum praealtum A. DC.	CRS	**pimil majben**	100
Smilacina amoena H.L. Wendl.	LIL	**ch'ib wamal**	60
Phoradendron falcatum (Cham. & Schltdl.) Trel.	VIS	**yijkats' te'**	80
Isochilus auramtiacum	ORC	**pech'e ch'e**	10
Elaphoglossum sp.	PTR	**takin tsib**	60
Deppea grandifolia	RUB	**tsajal te'**	30
Symplocos limoncillo Bonpl.	SYM	**tsuj te'**	40

Fig. 5.4 Voucher specimen of *Aecmea nudicaulis* used in study

Fig. 5.5 Voucher specimen of *Sedum praealtum* used in study

Fig. 5.6 Voucher specimen of *Smilacina amoena* used in study

Fig. 5.7 Voucher specimen of *Phoradendron falcatum* used in study

Fig. 5.8 Voucher specimen of *Isochilus auramtiacum* used in study

Fig. 5.9 Voucher specimen of *Elelphoglossum* sp. used in study

Fig. 5.10 Voucher specimen of *Deppea grandifolia* used in study

Fig. 5.11 Voucher specimen of *Symplocos limoncillo* used in study

lower on others. For example, only one informant reported a medicinal use for *Isochilus auramtiacum*. Interestingly, none of the informants responded affirmatively when asked if they had ever actually used any of these plants as medicinals.

The use recall interviews found 59 uses of medicinal plants (Fig. 5.12). The frequency with which medicinal plants were collected from particular habitats closely matches the results from the larger study discussed in Chap. 3. Over half of the time (52%) medicinal plants were collected from the highly disturbed areas of *akil* (grassland, trailside), *pat na* (area around house) or were cultivated. The frequency with which plants were collected from the various successional stages also matches the results of the larger study. Relatively few plants were collected in the very early successional stages but later stages (3–7 year and 6–12) were utilized 20% of the time. Young secondary forest (*te'tikil*) was utilized 13% of the time. There was only

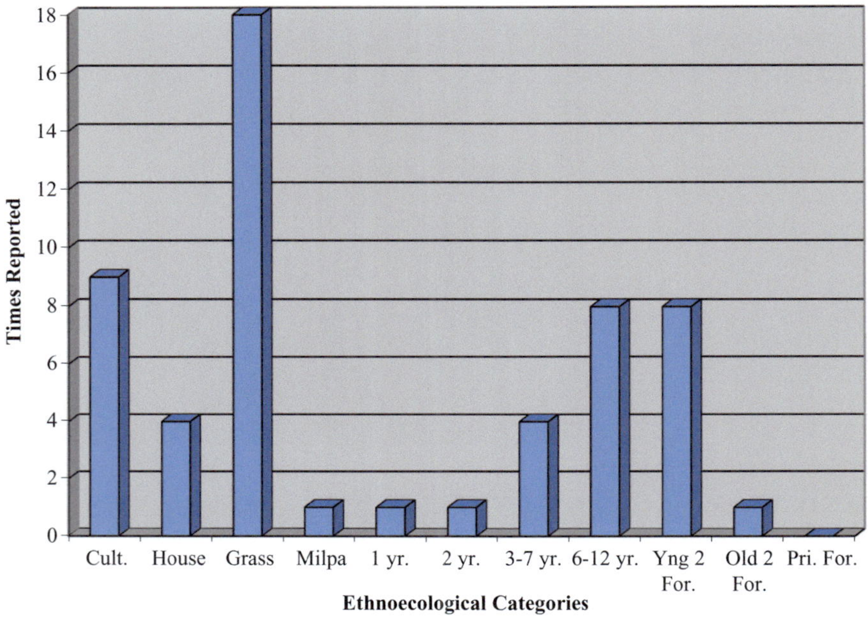

Fig. 5.12 Habitats where medicinal plants obtained by Tzeltal Maya living adjacent to mature forest

one reported procurement of a medicinal plant in old secondary forest and there was no use of the mature forest.

The average length of time to obtain a medicinal plant was 50 min. This is a significantly longer than the 16 min for the main study population. However, there were two instances were informants reported spending 5 h to obtain medicinal plants from hot country communities. If these outliers are removed the average time is 19 min. The total reported relative abundance of medicinal plants was 4.3 on a scale from one to five. The total reported relative strength (*yip)* of the medicinal plants was 4.4 on a scale from one to five.

5.4 Discussion

The close correspondence between the results of this chapter and that of Chap. 3 suggest that lack of availability and access to mature forests is not the reason that they are unutilized for medicinal plant procurement. Despite living within a kilometer of mature forests, none of the informants in this study reported use of mature forest, virtually no use of old secondary forest, and relatively little use of young secondary forest.

What is the significance of these results for the assumption that lowland mature tropical forest is the most important habitat for indigenous peoples to obtain medicinal plants? Montane tropical mature forest is, in many ways, significantly different than lowland mature forest. However, it is similar in that it is characterized by multi-layered hardwood species, high overall diversity and high in endemic species. Also, the fact that forests in general are under-utilized for medicinal plants suggests that there may be processes involved that are not particular to Highland Chiapas and that these results may have significance in other tropical areas as well. In the next chapter (Chap. 6) I present some possible explanations for these results based on theory and evidence from human ecology and biochemical ecology.

References

Abelson PH (1990) Medicine from plants. Science 247:513

Alcorn JB (1984) Huastec Maya ethnobotany. University of Texas Press, Austin

Balick MJ (1990) Ethnobotany and the identification of therapeutic agents from the rainforest. In: Chadwick DJ, Marsh J (eds) Bioactive compounds from plants. Wiley, Chicester, West Sussex, pp 21–23

Balick MJ, Arvigo R (2015) Messages from the Gods: a guide to the useful plants of Belize. Oxford University Press, New York

Balick MJ, Mendelsohn R (1992) Assessing the economic value of traditional medicines from tropical rain forests. Conserv Biol 6(1):128–130

Berlin B, Breedlove DE, Raven P (1974) Principles of Tzeltal plant classification. An introduction to the botanical ethnography of a Mayan-speaking people of the Highlands of Chiapas. Academic, New York

Booth W (1987) Combing the earth for cures to cancer, AIDS. Science 237:969–970

Bubb P (1991) The current situation of the cloud forest in Northern Chiapas, Mexico. Report to Ecósfera, Pronatura, The Linnean Society of London, and the Fauna and Flora Preservation Society

Calderon AI et al (2001) Forest plot as a tool to demonstrate the pharmaceutical potential of plants in a tropical forest of Panama. Econ Bot 54(3):278–294

Caniago I, Siebert SF (1998) Medicinal plant ecology, knowledge and conservation in Kalimantan, Indonesia. Econ Bot 52(1):229–250

Casagrande D (2004) Conceptions of primary forest in a Tzeltal Maya community: implications for conservation. Hum Organ 63(2):189–202

Christensen J (1999) Scientist at Work: Mark Plotkin; A romance with a rain forest and its elusive miracles. New York Times. November 30: Sec. F, p 3, Column 1

Cox P, Balick M (1994) The ethnobotanical approach to drug discovery. Sci Am 270(6):82–87

D'Andrade R (1995) The development of cognitive anthropology. Cambridge University Press, Cambridge

Earth Talk (2017) Tropical rainforests: nature's medicine cabinet. https://www.thoughtco.com/tropical-rainforests-natures-medicine-cabinet-1204030. Updated 9/13/17. Website accessed 9/22/17

Farnsworth NR, Soejarto DD (1991) Global importance of medicinal plants. In: Akerele O, Heywood V, Synge H (eds) The conservation of medicinal plants. Cambridge University Press, Cambridge, pp 25–52

Frei B, Sticher O, Heinrich M (2000) Zapotec and Mixe use of tropical habitats for securing medicinal plants in Mexico. Econ Bot 54:73–81

Jackson DD (1989) Searching for medicinal wealth in Amazonia. Smithsonian 19:95–101

Kempton W, Boster JS, Hartley JA (1995) Environmental values in American culture. MIT Press, Cambridge

Kuhn T (1970) The structure of scientific revolutions, 2nd edn. University of Chicago Press, Chicago
Mendelshohn R, Balick MJ (1995) The value of undiscovered pharmaceuticals in tropical forests. Econ Bot 49(2):223–228
Plotkin MJ (1991) Traditional knowledge of medicinal plants—the search for new jungle medicines. In: Akerele O, Heywood V, Synge H (eds) The conservation of medicinal plants. Cambridge University Press, Cambridge, pp 53–64
Plotkin MJ (1993) Tales of a Shaman's apprentice. Viking Press, New York
Posey DA (1984) A preliminary report on diversified management of tropical forest by the Kayapó Indians of the Brazilian Amazon. In: Prance GT, Kallunki JA (eds) Ethnobotany in the neotropics. New York Botanical Garden, New York, pp 112–126
Principe PP (1991) Valuing the biodiversity of medicinal plants. In: Akerele O, Heywood V, Synge H (eds) The conservation of medicinal plants. Cambridge University Press, Cambridge, pp 79–124
Schultes RE, von Reis S (1995) Ethnobotany: evolution of a discipline. Timber Press, Portland
Skirycz A, Kierszniowska S, Méret M, Willmitzer L, Tzotzos G (2016) Medicinal bioprospecting of the Amazon rainforest: a modern Eldorado? Trends Biotechnol 34(10):781–790
Stepp JR (2004) The role of weeds as sources of pharmaceuticals. J Ethnopharmacol 92:163–166
Stepp JR, Moerman DE (2001) The importance of weeds in ethnopharmacology. J Ethnopharmacol 75:19–23
Torrence PF (2012) Owed to nature: medicines from tropical forests. https://www.rainforesttrust.org/news/owed-to-nature-medicines-from-tropical-forests/. Website accessed 08/04/17
Voeks RA (1996) Tropical forest healers and habitat preference. Econ Bot 50:381–400
Voeks RA (2004) Disturbance pharmacopeias: medicine and myth from the humid tropics. Ann Assoc Am Geogr 94(4):868–888
Voeks R (2018) The ethnobotany of Eden: rethinking the jungle medicine narrative. University of Chicago Press, Chicago

Chapter 6
Medicinal Plants, Human Ecology and Biochemical Ecology

It is clear that the Tzeltal Maya overwhelmingly favor so called "disturbed" habitats for medicinal plants. In this regard, they show similarities with other sociolinguistic groups across the tropics and elsewhere that tend to favor human modified environments and herbaceous plants for medicines (Voeks 2004; Farias de Santana et al. 2016; Albuquerque et al. 2007). Despite rather fanciful popular notions of where small-scale societies find phytomedicines, the evidence to date clearly suggests that closed canopy mature forests are the least favored habitats. As I previously mentioned, this in no way should diminish the importance of these habitats for conservation, as they provide a range of ecosystem services (Costanza et al. 2007). It also does not dismiss the assumption that deep within the forest lie bioactive compounds in plants that can be turned into important and useful medicines. Nature is complex. Patterns and general trends do not equal immutable laws. However, it does suggest that discovery of such compounds will not come from ethnobotanical knowledge gained from trial and error but simply getting lucky through analysis and bioassays of previously untested or little explored plant species. But back to the matter at hand, what might explain why human modified landscapes are utilized more often for procurement of medicinal plants?

6.1 Human Ecology and Medicinal Plants

When one looks at non-specialized healing by the general population a few aspects become apparent. As previously discussed, there are numerous anthropological studies that suggest disturbed areas are preferred habitats for obtaining medicinal plants. Widely utilized medicinal plants need to be abundant and accessible (Stepp and Moerman 2001). Rare plants are not often found in medicinal floras because they either quickly become extirpated or are brought under cultivation, and thus cease being rare. Plants that take several days to find and collect are not very beneficial when one is sick. Thus plants that are close by will be preferred (Stepp and

J. R. Stepp, *Ethnoecology and Medicinal Plants of the Highland Maya*, Ethnobiology, https://doi.org/10.1007/978-3-319-69315-6_6

Moerman 2001). Traditional peoples, even those *of* the primary forest, generally do not live *in* the forest but rather on its edges, in riverine environments or in cleared areas (see, as examples, Rappaport 1984; Descola 1993; Chagnon 1997). This is certainly true with regards to horticultural and agricultural societies but perhaps less so for foraging societies and, of course, there are always exceptions (e.g. Shepard 1999). This issue of accessibility may be part of the reason that plants from disturbed areas have such significant representation in medicinal floras.

But as demonstrated herein, people will, occasionally, walk a long distance to obtain a medicinal plant if it is the only plant known to treat a certain condition. Thus a medicinal flora needs to be efficacious and a medicinal flora needs to be biologically active, for the most part. Although efficacy is a cultural construct (Moerman 2002), it is a cultural construct with a broad biological foundation. Plants that are not regarded as efficacious will be discarded for ones that are. There is abundant scientific literature attesting to the bioactivity of many of the most utilized Tzeltal Maya medicinal plants. It is highly unlikely that the continued use of these plants is due only to their symbolic value. However, it is important to note that placebo effects *are* present in treatments with medicinal plants, just as they are present in pharmaceutical treatments. Placebo effects are a complex cultural phenomenon and are likely present in all healing systems (Moerman 2000, 2002). This does not negate the demonstrated potential for medicinal plants to be bioactive and efficacious.

Once adopted, ethnopharmacopeias show remarkable persistence over time. Perhaps the most striking example comes from the Monte Verde site in Chile that gained notoriety when independently verified dates were established for human presence at the site definitively at least 13,000 years ago (Dillehay 1989) with additional evidence pointing to human presence between at least ~18,500 and 14,500 BP (Dillehay et al. 2015). This finding predated the Clovis site in New Mexico which had previously been considered the oldest known site of definite human occupation in the New World and has led to subsequent upheaval with our understandings of the peopling of the New World. At the site, archaeologists found remnants of 23 species of plants still known as medicinals by the present day indigenous Mapuche living in the area. Of the 23 plants, 10 have no other uses other than as medicines. Four of the species did not grow in the region, meaning that they had to have been brought in, a strong indication of their ethnobotanical significance. This also indicates that there were exchange networks between several different indigenous groups dating back to at least 14,500 BP (Dillehay et al. 2015). While it may be unlikely that the use of these plants as medicines represents a continuous transmission of knowledge over thousands of years by one sociolinguistic group, it nevertheless demonstrates persistence, human sociality and intergroup sharing and transmission of culturally important information and perhaps repeated re-discovery of efficacious phytomedicinals.

The argument thus far is for traditional peoples with a medicinal plants commons, where there is more or less egalitarian access to the resources. This scenario may become less common with increasing market integration of so many small-scale societies. However, even in a market-based economy, the above reasoning hold true. If anything, market incentives would accelerate the process of cultivating medicinal plants with limited distribution.

6.2 Biochemical Ecology and Medicinal Plants

Findings from the interdisciplinary field of biochemical ecology (or ecological biochemistry) can also do much to illuminate our understanding of why humans behave the way do with regards to medicinal plants. Ecological and biochemical theories and findings in relation to distribution and concentration of plant secondary compounds provide tremendous support for why plants from disturbed areas would tend to be more bioactive. These qualitative (*sensu* Feeny 1976) low molecular weight compounds are can serve as medicinals for humans (Kinghorn 1994).

Secondary compounds in plants are important for a variety of ecological functions. Chief among these are as insect and animal attractants for pollination and seed dispersal; allelopathy, where they inhibit germination and growth of other plants; and, as chemical defense against herbivory (Harborne 1993). The International Allelopathy Society defines it as "Any process involving the secondary metabolites produced by plants, microorganisms, viruses, and fungi that influence the growth and development of agricultural and biological system (excluding animals), including positive and negative effects" (Torres et al. 1996).

With regards to allelopathy, dozens of species of plants have been shown to interfere with crops through secondary compounds (Putnam 1985; Latif et al. 2017) although the number is undoubtedly much higher. Because allelopathy usually occurs through the complex chemical matrix of the soil it can be difficult to conclusively show a causal relationship (Zimdahl 2018) and their remains much work to be done in this area (cf. Rice 1984). Perhaps not surprisingly, some of the most important families for medicinal plants such as Compositae, are also known for containing many species that exhibit allelopathic potentials (Chon and Nelson 2010).

Investigations into plant anti-herbivore defense are further developed although still evolving. Scientists have long pondered what Feeny (1976) termed a conspicuous non-event on the planet: given the enormous destructive potential of herbivorous insects, why have they failed to prevent plants from dominating most of the terrestrial surface of the planet. In short, why is the planet so green? A response is neatly provided by Janzen (1975), 'The world is not colored green to the herbivore's eyes, but rather is painted morphine, L-DOPA, calcium oxalate, cannabinol, caffeine, mustard oil, strychnine, rotenone, etc."

Yet anti-herbivory secondary compounds in plants are not equally distributed across the landscape. Plants vary enormously in their strategies to prevent herbivory. Yet researchers have noted some patterns. Two major theories of plant defense, apparency theory (Feeny 1976; Rhoades and Cates 1976) and resource availability theory (Coley et al. 1985) are worth exploring with regards to medicinal plants and I discuss these two theories in more detail below. I should note that there has also been considerable work done with regards to the role of environmental constraints and carbon/nutrient balance that is less relevant to this discussion but also does not necessarily contradict the findings in apparency theory and resource availability theory (Stamp 2003; Frew et al. 2016). Both apparency theory and resource availability theory support the hypothesis that vegetation in disturbed areas would tend to be high in bioactive secondary compounds (Stepp and Moerman 2001; Stepp 2004). These two theories are complementary to each other and share some predictions such as the prevalence of bioactive compounds in early successional species and digestibility reducers in late successional and climax species (Howe and Westley 1988). Like all theories there are exceptions to the rule. Both theories have been criticized because of these exceptions. This points to the difficulty of developing a general theory of plant defense, given the wide range of strategies that plants engage in and the wide range of habitats in which they live. The explanatory value of both theories is greatest when one looks at general cases. Also, in as much as these theories dichotomize, they provide a model of nature. The reality is probably closer to a continuum of defense strategies with significant clustering into two overall patterns.

6.2.1 Apparency Theory

Feeny (1976) suggests that there are two main types of antiherbivory chemical defense strategies for plants depending on their degree of apparency (i.e. how likely they are to be discovered by herbivores). Unlike animals, plants can not flee predation and require defenses. First, there are metabolically inactive, *quantitative defenses* such as tannins and lignins of high molecular weight. These compounds have been shown to reduce digestibility but do not act as biological toxins (Feeny 1976). These generalized and oftentimes structural defenses, are advantageous for plants that are easily found. This correlates with the lifespan of the plant. Feeny notes that trees and shrubs are highly apparent along with perennial grasses. Grasses rely on structural defenses such as silica, or they simply grow back after they are consumed by herbivores. This may explain the significant under-representation of medicinals from the Poaceae family in the Highland Chiapas ethnopharmacopeia and ethnopharmacopeias worldwide. Since grasses do not usually rely on qualitative defense compounds they rarely have bioactive secondary compounds. Consequently they are rarely used as medicinals, and instead are the single most important source of human food plants (wheat, maize, rice, barley, sorghum, oats, rye, sugar cane, etc.)

The second type of strategy relies on *qualitative* defenses, low molecular weight secondary compounds such as alkaloids, cardiac glycosides or terpenoids. These compounds are toxic at low concentrations (usually <1% dry weight) and highly biologically active (Feeny 1976; Coley et al. 1985). Ephemeral, successional or *r*-selected species, meaning species that are opportunistic, rapidly colonize an area, and are short-lived tend to rely on these sorts of toxic chemical defenses (Rhoades and Cates 1976; Abe and Higashi 1991). These plants are less likely to be discovered by insects and thus less in need of generalized defenses that require a significant metabolic cost and investment. Qualitative defenses usually serve the plants well against herbivory because their short life span makes it less likely that a population of insects will evolve ways to detoxify the compounds. Of course, this is not always the case and there is abundant literature noting the ability of specialized herbivores to bypass plant qualitative defense compounds (Harborne 1993).

Apparency theory can also be applied to predictions about specific parts of plants. McKey (1979) hypothesizes that in the same way that plants with long life spans invest in quantitative defenses, so will the longer living parts of plants. He suggests that quantitative compounds are more likely to show up in woody tissues and mature leaves while new leaves are more likely to contain qualitative compounds. This may explain why the Tzeltal usually use fresh leaf material and young stems to prepare treatments rather than more woody parts of plants.

6.2.2 Resource Availability Theory

In many ways, apparency theory is a theory developed from the perspective of the animal in that it implies that herbivory is the selection pressure that allows for the evolution of defense compounds. A more recent body of theory termed resource availability theory emphasizes the plant perspective. In some regards, it is a more robust theory in terms of explaining certain patterns of plant defense (Howe and Westley 1988).

Resource availability theory (Coley et al. 1985) derives from an assumption that the availability of resources for plants to replace parts lost to herbivory is the primary determinant in the type of defense. It was similarly suggested as the "rich man spendthrift hypothesis" by Janzen (1985: 210) when he stated that "the harder it is for a plant to replace lost tissues, taken by an herbivore, the more intense should selection be for those traits that prevent herbivory."

There are two main variables affecting growth potential that determine which defense strategy plants will engage in: (1) growth rate and lifespan; (2) resource availability (including light) and resource type. With regards to growth rate and lifespan, Coley et al. (1985) state that fast-growing species tend to invest in mobile (qualitative *sensu* Feeny) defense compounds rather than immobile (quantitative *sensu* Feeny) compounds. They also demonstrate a positive correlation between short-living plants and a reliance on mobile compounds. Further support for the relationship between lifespan and type of defense is provided by the fact that all over the world, alkaloids are twice as likely to appear in annuals than perennials (Levin 1976). Leaf life is also an important predictor of defense strategy (Coley 1988). Plants with long-lived leaves rely on immobile defenses while plants with short-lived leaves invest in toxic compounds.

Resource availability theory also predicts that plants that are growing in conditions with abundant light and fertile soil will invest in mobile defenses because they can more readily replace lost leaves. The metabolic cost is less to develop mobile defenses. When resources are not as available, each leaf becomes valuable to the plant. Thus, the plant will tend to invest in immobile defense which, while more expensive initially, protect the leaf better over an extended period of time.

Both apparency theory and resource availability theory suggest a biochemical explanation for why humans would prefer disturbed habitats for medicinal plants. They predict that herbaceous plants in early successional habitats are more likely to contain bioactive compounds than trees and shrubs in mature forest. They demonstrate that the characteristics of weeds in particular make them more likely to contain bioactive compounds. These are the same types of plants from the same types of habitats that the Tzeltal use as phytomedicinals.

References

Abe T, Higashi M (1991) Cellulose centered perspective on terrestrial community structure. Oikos 60:127–133

Albuquerque UP, de Medeiros PM, de Almeida AL, Monteiro JM, Neto EM, de Melo JG, dos Santos JP (2007) Medicinal plants of the caatinga (semi-arid) vegetation of NE Brazil: a quantitative approach. J Ethnopharmacol 114(3):325–354

Chagnon NA (1997) Yanomamö, 5th edn. Harcourt Brace, Fort Worth

Chon SU, Nelson CJ (2010) Allelopathy in compositae plants: a review. Agron Sustain Dev 30:349–358

Coley PD (1988) Effects of plant growth rate and leaf lifetime on the amount and type of anti-herbivore defense. Oecologia 74:531–536

Coley PD, Bryant JP, Chapin FS (1985) Resource availability and plant anti-herbivore defense. Science 230:895–899

Costanza R, Fisher B, Mulder K, Liu S, Christopher T (2007) Biodiversity and ecosystem services: a multi-scale empirical study of the relationship between species richness and net primary production. Ecol Econ 61(2–3):478–491

Descola P (1993) The spears of twilight: life and death in the Amazon jungle. The New Press, New York

Dillehay TD (1989) Monte Verde: a late Pleistocene settlement in Chile, vol 1. Smithsonian Institution Press, Washington, DC

Dillehay TD, Ocampo C, Saavedra J, Sawakuchi AO, Vega RM, Pino M et al (2015) New archaeological evidence for an early human presence at Monte Verde, Chile. PLoS One 10(11):e0141923

Farias de Santana B, Voeks RA, Funch LS (2016) Ethnomedicinal survey of a maroon community in Brazil's Atlantic tropical forest. J Ethnopharmacol 181:37–49

Feeny PP (1976) Plant apparency and chemical defense. In: Wallace JW, Mansell RL (eds) Recent advances in phytochemistry. Plenum, New York, pp 1–40

Frew A, Powell JR, Sallam N et al (2016) Trade-offs between silicon and phenolic defenses may explain enhanced performance of root herbivores on phenolic-rich plants. J Chem Ecol 42:768–771

Harborne JB (1993) Introduction to ecological biochemistry, 4th edn. Harcourt Brace and Company, London

Howe HF, Westley LC (1988) Ecological relationships of plants and animals. Oxford University Press, Oxford

Janzen DH (1975) Ecology of plants in the tropics. Edward Arnold Publishers, London

Janzen DH (1985) Plant defences against animals in the Amazonian rainforest. In: Prance GT, Lovejoy TE (eds) Amazonia. Pergamon Press, Oxford

Kinghorn AD (1994) The discovery of drugs from higher plants. In: Gullo VP (ed) The discovery of natural products with therapeutic potential. Butterworth-Heinemann, Boston, pp 81–108

Latif S, Chiapusio G, Weston LA (2017) Allelopathy and the role of allelochemicals in plant defence. In: Becard G (ed) Advances in botanical research, vol 82. Academic, New York, pp 19–54

Levin DA (1976) Alkaloid-bearing plants: an ecogeographic perspective. Am Nat 110:261–284

McKey D (1979) The distribution of secondary compounds within plants. In: Rosenthal GA, Janzen DH (eds) Herbivores: their interactions with secondary plant compounds. Academic, New York

Moerman DE (2000) Cultural variations in the placebo effect. Med Anthropol Q 14:51–72

Moerman DE (2002) Meaning, medicine and the 'placebo effect'. Cambridge University Press, New York

Putnam AR (1985) Weed allelopathy. In: Duke SO (ed) Weed physiology, vol I: reproduction and ecophysiology. CRC Press, Boca Raton, FL, pp 131–155

Rappaport RA (1984) Pigs for the ancestors, 2nd edn. Yale University Press, New Haven

Rhoades DF, Cates RG (1976) Toward a general theory of plant antiherbivore chemistry. In: Wallace JW, Mansell RL (eds) Recent advances in phytochemistry, vol 10. Plenum, New York, pp 168–213

Rice EL (1984) Allelopathy. Academic, New York

Shepard GH (1999) Pharmacognosy and the senses in two Amazonian societies. University of California, Berkeley. PhD dissertation

Stamp N (2003) Out of the quagmire of plant defense hypotheses. Q Rev Biol 78(1):23–55

Stepp JR (2004) The role of weeds as sources of pharmaceuticals. J Ethnopharmacol 92:163–166

Stepp JR, Moerman DE (2001) The importance of weeds in ethnopharmacology. J Ethnopharmacol 75:19–23

Torres A, Olivia RM, Castellan D, Cross P (1996) A science of the future. In: Proceedings of the 1st World Congress on Allelopathy, University of Cadiz, Cadiz, Spain

Voeks RA (2004) Disturbance pharmacopeias: medicine and myth from the humid tropics. Ann Assoc Am Geogr 94(4):868–888

Zimdahl RL (2018) Fundamentals of weed science, 5th edn. Academic, San Diego, CA

Chapter 7
Medicinal Plant Ethnoecology and Climate Change: Implications for Conservation, Health and the Environment

Knowing where medicinal plants are actually procured can allow for improved management and access to these crucial resources. It also allows for mitigation in the face of environmental change, especially climate change. The work presented in previous chapters is concordant with research on medical ethnobotany with the Highland Maya demonstrating that medicinal plants play an overwhelmingly primary role in the daily health care of the Highland Maya (Berlin et al. 1974; Berlin and Berlin 1996). Despite having access to clinics in most of the communities, people continue to self-administer treatments derived from medicinal plants and rely on clinics on rare occasions for very serious conditions. Several factors are at work that, despite external pressures that might contribute to knowledge loss such as increased market integration and mobility, have prevented degradation of traditional ecological knowledge about medicinal plants. The population of Tzeltal Maya speakers is large and growing, with at least 450,000 speakers in the region which represents a doubling in size since 1980 (INEGI 2011). Illness rates and marginality indices remain high in most Tzeltal communities, making knowledge of medicinal plants not just practical but a matter of survival. While western medical clinics are available to some degree, cultural preferences strongly lean towards self-medicating with medicinal plants that are freely available and, in most cases, nearby. Recall that for the Tzeltal Maya, the vast majority of medicinal plant procurement is conducted within a 2-km radius around each household and plants with limited ecological distribution are usually utilized only in communities where they are readily available. The average time spent in procurement of medicinal plants for a specific and discrete treatment is 16 min. This correlates with data demonstrating that a 1 km radius around a given household is the primary area for obtaining medicinal plants. Eighty-six percent of medicinal plants are obtained in managed areas or areas of secondary vegetation (non-forest). Fourteen percent are obtained in young secondary forest. Research on the ethnoecological distribution of 203 plants with the highest consensus regarding their status as medicinals show that early successional stages are the most common habitats for medicinal plants and that distribution falls off sharply in forested areas. No medicinal plants were obtained in areas

J. R. Stepp, *Ethnoecology and Medicinal Plants of the Highland Maya*, Ethnobiology, https://doi.org/10.1007/978-3-319-69315-6_7

of old secondary or primary forest. Even people living adjacent to primary forest (within a 1 km radius of primary forest) could not recall ever obtaining medicinal plants in the primary forest for any illness, although they could identify a small set of eight plants that were previously identified as primary forest obligate medicinal plants. The ratio of use between secondary non-forest habitats and secondary forest was comparable to that found in communities that are not near primary forests.

7.1 Social and Environmental Change

Highland Maya medicinal plant ethnoecology has proven to be dynamic and resilient over millennia. Tremendous social and environmental change has occurred at various times in the past with notable events including the depopulation of urban areas (the so-called collapse of ancient Maya civilization), European conquest, the *encomienda* system, Mexican independence, the Mexican revolution (even though its reforms largely failed to reach Chiapas), huge population increases beginning at the latter part of the twentieth century and the Zapatista uprising. Through it all, Highland Maya have continued to rely on medicinal plants for their primary healthcare.

The latest disruptive event is climate change. There is some evidence that it is already having an effect in montane regions at a greater rate than lowland areas and Chiapas is no exception (Rangwala and Miller 2012; Loarie et al. 2009). Before discussing some potential impacts though, it is important to look through the long lens of history. It shows us time and time again that, in the face of dire circumstances, the Highland Maya have persevered and survived. Within this context, climate change is a serious threat but not an apocalyptic one (cf. Barnett and Adger 2007).

Climate prediction is an inexact science but current predictions for Chiapas indicate that temperatures will increase by 1.6 °C by 2030 (World Bank 2018). Longer-term predictions indicate a 2–3 °C rise in the twenty-first century under a medium high emission scenario (Conde et al. 2011). This brings with it an increase of heat waves lasting on average 10–20 days longer. Annual precipitation is expected to decline by 2030 in a range between −6 and −53 mm (World Bank 2018). Despite an overall annual decline, the region will see more intense rainfall events but a shorter rainy season and thus, a longer dry season.

Agriculture in Chiapas is already seeing some effects of climate change and can provide some insights for how medicinal plants are being affected (Donatti et al. 2018). For example, smallholder coffee producers have noted a decrease in yield (Ruiz Meza 2015). The elevational range in which coffee can be produced is expected to shift upwards by 100 m for every 0.8 °C increase in temperature (Schroth et al. 2009). Warmer temperatures will lead to increase predation by insects and more favorable conditions for other pathogens (Bunn et al. 2018). Maize production will also be impacted by climate change. Bellon et al. (2011) explored traditional maize seed-saving systems in nearby Central Mexico and concluded that farmers in highland regions would be the most impacted and would likely need to acquire new seed outside their traditional geographical ranges. There is a strong overlap between food plants and medicinal plants (and sometimes no distinction is made between the two categories). Food plants receive a great deal of attention to their production compared to medicinal plants and may serve as proxies for potential impacts of climate change on medicinal plants.

Based on what is now known about Highland Maya medicinal plant ethnoecology, what might be the impacts of climate change? The fact that the most utilized medicinal plant species are abundant and found nearby suggests that climate change will have little impact on the availability of those plants for particular communities that are currently utilizing them. In marginal areas where their distribution is more limited, they may become hard to find but it would require massive ecological disruption for them to be extirpated from the region.

The broader issue then, is the impact of climate change on the approximately 200 species that are regularly procured and the approximately 1300 other species that are known but less utilized. Medicinal plant species in these categories that are abundant now tend to be that way because of their wide distribution across successional stages and major ecological zones. They should also only see minor impacts from climate change. Much like the predicted rise of the elevation in which coffee is produced (Schroth et al. 2009), the medicinal flora may also migrate upwards in elevation. The highest elevations in Chiapas are where primary cloud forests remain and are areas where medicinal plants are not generally found. Some species may not find appropriate habitats at these higher elevations as a result. Confounding the situation is figuring out what will happen to these primary forests (Ponce-Reyes et al. 2012; Rojas-Soto et al. 2012) and other forested areas (Gómez-Mendoza and Arriaga 2007) in the face of climate change. Loss of these forests will be unfortunate and should be avoided but it may create additional successional habitats for some medicinal plant species. However, even that potential minor benefit will be offset by impacts on human health due to erosion and increased flooding risks from higher elevation forest loss.

Obscure and less utilized species are likely to see more impacts from climate change. They are less utilized precisely because they are less abundant and found in fewer habitats. Some of these plant species may disappear, or, as they become increasingly scarce, knowledge of their medicinal importance may be lost. The overall impact on human health is debatable. Most of these species are interchangeable, at least in relation to the conditions they treat. Put another way, the number of ethnomedical conditions present is far fewer than the number of medicinal pant species that treat them. If one plant becomes less available, its likely that another species would take its place in the medicinal flora. Already, substitution is common, with individuals regularly making choices as to which plant species to treat a particular condition based on both availability, personal preference and by extension, community preference. Communities are adapted to particular elevations in their selection of preferred medicinal plants and individuals will rarely travel far afield for a medicinal plant when another one nearby could substitute.

Previous work (Stepp 2017) demonstrated that there is diversity in knowledge throughout the region with the variation increasing significantly at the inter-community level (i.e. between communities). Intra ecological zone variation is also high, even when communities are adjacent. This all indicates that most transmission of traditional plant knowledge occurs at the household and community level. We have all experienced an analogous situation when faced with a common illness such as a cold. Upon displaying symptoms, an individual may receive advice within his household from his immediate family as to how to treat it. They may then receive additional advice from their work, or school or other community locations outside of the home. Rarely (at least until the rise of the internet) would one receive advice (knowledge) outside of their community.

The Highland Maya are much like the rest of us then in how they learn about medicinal plants, or at least this is what is suggested by the variation data (Stepp 2017). This scenario of widespread sharing within a community with limited sharing beyond communities has served the Highland Maya well for millennia. However, in the face of rapid environmental and climate change, there is a need for increased sharing of medicinal plant knowledge between communities. For example, a community at a different elevation may know plants that will fare better or that will migrate to new elevations with changes in climate. This knowledge could be made available to affected communities. We have had some success in the past with bringing communities together to discuss medicinal plants and share new knowledge. Highland Maya enjoy talking about plants and this willingness to share new information to neighboring communities could help to mitigate the impacts of climate change on medicinal plants.

So far, this discussion has focused primarily on access and, by extension, yield but plant quality (as defined by secondary chemistry that contains health-related bioactive compounds that provide efficacy for medicinal plants) is also important to consider (Ahmed and Stepp 2016). We know, for example, that higher CO_2 levels can impact nutrient quality of plants (Myers et al. 2014). How might particular medicinal plant species respond to climate change in terms of their chemistry? This is a very complex and little explored question. We know that increased predation by insects can lead to increased production of defense compounds (see Chap. 6 for a full discussion), and insect predation will increase in a warming climate. A simplistic answer then would be that medicinal plants might show increased bioactivity. However, the devil is in the details and so little is understood about how secondary chemistry in individual plant species responds to environmental change and variation that it would be irresponsible to make such claims. What can be asserted is that climate change will affect medicinal plant quality and chemistry in addition to plant abundance and yield and deserves further investigation.

7.2 Implications for Conservation

Beyond climate change and before presenting some final remarks, I would like to discuss the implications of this work for conservation, both in general and in the Highlands of Chiapas. The intention of the research is not to attack economic arguments for preservation of primary forest despite results demonstrating the non-use of primary forest for medicinals. Yet I do feel that making such arguments is dangerous even when there is good supporting data. For example, if the main reason for preservation of primary forest is for their value in providing novel compounds for pharmaceuticals, they would only remain valuable until the compounds could be discovered and synthesized. Economic arguments are subject to change and fluctuation in markets and technology in a way that intrinsic arguments for conservation are immune. While economic arguments are useful in reaching a broad audience with a conservation message, my opinion is that they should always remain secondary to arguments based on the intrinsic value of primary forests for humans.

Obviously primary forest in the Highlands of Chiapas is not going to be protected for its medicinal value. But there are other aspects of conservation apart from habitat conservation such as medicinal plant conservation, regardless of the habitats from which they come. While medicinal plants in the Highlands taken as a whole are not in any apparent danger of localized extinction, some species, by virtue of growing during a narrow period of time in the successional process, might be considered potentially threatened. The Tzeltal Maya system of obtaining medicinal plants from nearby early successional habitats appears to have served the Tzeltal well until recent surges in the human population may be affecting the distribution and availability of some species. As discussed in Chap. 2, deforestation and reduced fallow times (due in part to lack of cultivable lands and the increased use of chemical fertilizers) are common processes throughout the Highland region. These changes could be impacting the distribution of the medicinal flora. As the fallow cycle is reduced, the amount of regeneration in secondary habitats and forests is similarly reduced. This, in turn, potentially affects the distribution of the medicinal plant species that show up later in succession. Meanwhile, cultivation of medicinal plant species has been almost non-existent until recently. There are encouraging signs as communities develop medicinal plant gardens (Fig. 7.1) and it could

Fig. 7.1 Women in Oxchuc Working in Their Community Garden

provide an infrastructure for the conservation of medicinal plants and the transmission of knowledge across communities and municipalities.

7.3 Concluding Remarks

It is clear from this and many other studies in Highland Chiapas that the Highland Maya have an intimate knowledge of their biophysical environment. The landscape that the Highland Maya live in is very much a human landscape and research should reflect the complex interactions taking place. Especially with regards to the so-called 'disturbance' vegetation, it is important to understand human modification of the biophysical environment in a non-pejorative manner. The Highland Chiapas 'disturbance' vegetation and the medicinal plants it provides is likely the result of thousands of years of human-plant interactions. Even in the face of widespread social and environmental change, it has persisted and will continue to persist into the future as long as people continue to be reliant on their local environment for their primary healthcare. An ethnoecological framework can provide an understanding of these interactions that takes into account both ecological and sociocultural factors. I conclude with a question: What is in fact, more impressive and exotic? A solitary healer trekking deep into the forest to find a magical cure? Or, a culture that develops an empirically based system of healing over millennia and by transmitting this knowledge orally successive generations are able to find effective treatment for a wide range of illnesses from a 'living pharmacy' (Fig. 7.2) that they create within their biophysical environment?

Fig. 7.2 Community of Balum Kanal gathered together to share medicinal plant knowledge

References

Ahmed S, Stepp JR (2016) Model for effects of climate change on food quality and security. Elementa 4:000092

Barnett J, Adger WN (2007) Climate change, human security and violent conflict. Polit Geogr 26(6):639–655

Bellon MR, Hodson D, Hellin J (2011) Assessing the vulnerability of traditional maize seed systems in Mexico to climate change. Proc Natl Acad Sci 108(33):13432–13437

Berlin EA, Berlin B (1996) Medical ethnobiology of the highland Maya of Chiapas, Mexico: the gastrointestinal diseases. Princeton University Press, Princeton, NJ

Berlin B, Breedlove DE, Raven P (1974) Principles of Tzeltal Plant Classification. An introduction to the botanical ethnography of a Mayan-speaking people of the Highlands of Chiapas. Academic, New York

Conde C, Estrada F, Martínez B, Sánchez O, Gay C (2011) Regional climate change scenarios for México. Atmósfera 24(1):125–140

Bunn C, Castro F, Lundy M (2018) The impact of climate change on coffee production in Central America. CGIAR, Rome

Donatti CI, Harvey CA, Martinez-Rodriguez MR, Vignola R, Rodriguez CM (2018) Vulnerability of smallholder farmers to climate change in Central America and Mexico: current knowledge and research gaps. Clim Dev:1–23

Gómez-Mendoza L, Arriaga L (2007) Modeling the effect of climate change on the distribution of Oak and Pine species of Mexico. Conserv Biol 21(6):1545–1555

INEGI (2011) Mexico population and housing census 2010. Instituto Nacional Estadística, Geográfica e Informática, Mexico City

Loarie SR, Duffy PB, Hamilton H, Asner GP, Field CB, Ackerly DD (2009) The velocity of climate change. Nature 462(7276):1052

Ponce-Reyes R, Reynoso-Rosales VH, Watson JE, VanDerWal J, Fuller RA, Pressey RL, Possingham HP (2012) Vulnerability of cloud forest reserves in Mexico to climate change. Nat Clim Chang 2(6):448

Rangwala I, Miller JR (2012) Climate change in mountains: a review of elevation-dependent warming and its possible causes. Clim Chang 114(3–4):527–547

Rojas-Soto OR, Sosa V, Ornelas JF (2012) Forecasting cloud forest in eastern and southern Mexico: conservation insights under future climate change scenarios. Biodivers Conserv 21(10):2671–2690

Ruiz Meza LE (2015) Adaptive capacity of small-scale coffee farmers to climate change impacts in the Soconusco Region of Chiapas, Mexico. Clim Dev 7(2):100–109

Schroth G, Laderach P, Dempewolf J, Philpott S, Haggar J, Eakin H, Castillejos T, Moreno JG, Pinto LS, Hernandez R, Eitzinger A (2009) Towards a climate change adaptation strategy for coffee communities and ecosystems in the Sierra Madre de Chiapas, Mexico. Mitig Adapt Strateg Glob Chang 14(7):605–625

Stepp JR (2017) Transmission, sharing and variation of medicinal plant knowledge and implications for health. In: Olson L, Stepp JR (eds) Plants & health: new perspectives on the human relationships to medicinal plants. Springer, New York, pp 163–170

The World Bank (2018) Climate change knowledge portal. http://sdwebx.worldbank.org/climateportal/index.cfm

Myers SS et al. (2014) Increasing CO_2 threatens human nutrition. Nature 510 (7503):139–142

Index

J. R. Stepp, *Ethnoecology and Medicinal Plants of the Highland Maya*, Ethnobiology, https://doi.org/10.1007/978-3-319-69315-6

MIX
Papier aus verantwortungsvollen Quellen
Paper from responsible sources
FSC® C105338

If you have any concerns about our products,
you can contact us on
ProductSafety@springernature.com

In case Publisher is established outside the EU,
the EU authorized representative is:
Springer Nature Customer Service Center GmbH
Europaplatz 3, 69115 Heidelberg, Germany

Printed by Libri Plureos GmbH
in Hamburg, Germany